Design and Destiny

Design and Destiny:
the making of the
TUCKER AUTOMOBILE

Philip S. Egan

ON THE MARK ● Orange

Published by
ON THE MARK Publications
P.O. Box 5276, Orange, California 92863-5276

Copyright © 1989 by Philip S. Egan

First Printing   April 1989
Fifth Printing   August 1999

Cover Photo of Tucker automobile No. 4, owned by Ray Burton.

Photo by Frank Capri

Printed in the United States of America
by Braun-Brumfield, Inc.
Acid-free paper

Library of Congress Cataloging-in-Publication Data:

Egan, Philip S., 1920-
  Design and destiny.

  Includes index.
  1. Tucker automobile—History.   I. Title.
TL215.T83E33   1989   629.2'222   89-90666
ISBN 0-924321-00-8

10 9 8 7 6 5

To Preston Tucker,
to those who helped him make the Tucker '48 possible,
and to the many who still carry the memory and message of the car,
this book is dedicated.

TABLE OF CONTENTS

ACKNOWLEDGMENTS

Research for this book sustained rapport with those people who shared in the excitement of the Tucker project with the author. Numerous contacts developed with others who were able to provide invaluable information. The responses of these helpful people were overwhelming in scope and a source of reassuring encouragement throughout the effort. In alphabetical order, they are:

Shelby C. Applegate,
  Applegate & Applegate,
    photographic archives
D. E. Barnett,
  Tatra owner/historian
Richard Burns Carson,
  author/historian
John C. Cermak,
  former secretary, Tucker Automobile Club of America
Virginia F. Egan,
  manuscript editor
Roy Nagel,
  Dubonnet photos
Beverly Ferriera,
  Tucker owner/historian
Ron Grantz, curator, Detroit Public Library,
  National Automotive History Collection

Richard E. Jones, founder,
    The Tucker Automobile Club of America
William J. Lewis, Research Historian,
    The Lewis Automotive History Collection
J. Gordon Lippincott, co-founder,
    J. Gordon Lippincott and Company (Lippincott & Margulies)
Gérard Lorieux, Technical Relations Director,
    Citroen, Neuilly-Sur-Seine, France
Tucker P. Madawick, FIDSA,
    project compatriot/historian
Cynthia Read Miller, Curator,
    The Henry Ford Museum
Bill Riemer,
    Tucker historian
John and Dorothy Ristine,
    Who loyally saved clay model sketches
Karel Rosenkranz, Director,
    Tatra Technical Museum, Koprivnice, Czechoslovakia
Dennis Schrimpf, Technical Editor,
    Old Cars Weekly
Budd Steinhilber, FIDSA,
    Project compatriot
Marilyn Story,
    Cadillac Motor Car Division, Detroit
Alex S. Tremulis, Chief Stylist,
    The Tucker Corporation
Read Viemeister, FIDSA,
    Project compatriot

PREFACE

The crowd, five thousand strong, rose to its feet and roared its approval. They were cheering the premier presentation of a car; an automobile design unique for 1947 and destined to remain unique for decades to come.

The story of the Tucker '48 automobile is one of the most colorful in the history of motor cars. Its creation began in the bright light which awakened mankind from the darkness of World War II.

A lone individual with a passion for improving automotive design gathered together a following for a remarkable enterprise. He was not famous to begin with, and is now hardly known at all, yet during the brief period of this saga, his intense belief in his goals attracted and inspired thousands of people.

He was Preston Thomas Tucker, an unlettered engineer intimately acquainted with some of the great names in automotive history. From singularly modest beginnings, he developed his experience into credentials for success, rising to unusual heights.

This is the story of the designing of his dream car, an automobile that proudly joins the ranks of advanced design concepts of automotive history; a car that still fascinates the public forty years later.

This is also the story of a young designer excited at being able to participate in a major automotive enterprise, one who knew Preston Tucker personally and who played a part in the creation of the Tucker '48.

I live in the future,
but I think of the past,
for I need the past
upon which to build
the future.

-PSE-

PART I
DESIGN

CHAPTER 1

AN AWAKENING

Picture a spring afternoon in 1936. As I walked from the Grosse Pointe High School bus stop on Jefferson Avenue in Detroit, Michigan, towards the Whittier Hotel, where I lived with my father, I was attracted to an unusual shape poised at the curb in front of the hotel's main entrance. It had the shape of an odd dark blue fish, dorsal fin and all. However, on closer inspection I confirmed that it was indeed an automobile, a most unusual automobile.

Then, as now, I was a car enthusiast. My father, Sidney B. Egan, began his career in the automobile industry before he switched to his forte, art, and became an advertising art director. He always retained his love of cars. When I was ten years old, he produced a rendering of the ideal futuristic vehicle for me. It was a vermilion watercolor done on black board, sleek, low and long, a portent of a Duesenberg of several years later. That drawing sparked in me a passion for drawing which became the substance of my livelihood. It also induced my own automotive interests.

My father excited the family with purchases of two Pierce Arrows in succession, lovely automobiles of majestic proportions. Attendance at auto shows was a regular family event, and cars a common topic of conversation at home. Quite a number of people in my childhood neighborhood of Evanston, Illinois (just north of Chicago) were fortunate enough to afford new Stutz Black Hawks, Cord L-29s and glistening Packards.

Art and design were a part of my young life. My mother, Alice
Brown, first met my father while they were students at the Art Institute in Chicago. Alice, an aspiring watercolorist, inspired my sister,
Margaret, five years my senior, to follow her path. Our family life was
expansive, branching out to include the world of music and the performing arts. By age eight, I could hum through Tschaikovsky's Nut-
cracker Suite. We attended first-run musicals at Chicago theaters, and
family friends included local luminaries of radio. We were not a
wealthy family. I think my father lived substantially beyond his means
at all times, but he possessed charm and talent beyond his material
worth.

Mother died in 1930 when I was nine years old. My father and sister continued to foster an appreciation for the arts. After Mother's
passing, we moved to New York, then to Cleveland, and back to New
York again as my father's career in advertising art developed.

There were many touches of the graphic arts in our homes. In our
apartment in Rye, New York, my father had the ceiling of the living
room painted flat black, a dramatic contrast with the chalk white walls
— psychologically lowering the ceiling two feet. Then there was the
expression of a pure, beautifully contemporary interior design in our
Manhattan apartment, with white lacquered cabinets, pewter lamps,
blond furniture, and massive glass accessories.

Margaret married and stayed in the east. My father and I then moved
to Detroit — another promotion — in 1934. Our shared interest in
music continued when he bought the Victor Library of music, hundreds
of the choicest classical albums. At age fourteen, alone much of the
time, I played these records and progressed from Tschaikovsky to
Stravinsky.

In the midst of this artistic atmosphere, cars remained an important
part of my life. In 1935 my father took delivery of a '35 Ford V-8
convertible sedan. A black beauty with tan rag top, chromium wheel
discs and white wall tires, the car was a magnet for attention. On several
occasions notes were placed on the windshield from people wanting to
buy it outright.

Thus I was well primed for the encounter with the unusual vehicle in
front of the Whittier Hotel, that bright spring day. I rushed up to our
apartment, grabbed my Argus 35mm camera, and tore back down-

stairs to take three shots of the car before it vanished from sight.

For 1936, this was a most remarkable automobile. The driver and front passenger were placed at the extreme forward end, with an aircraft style windshield and nose immediately in front of them. The outer shell of the car was completely different from anything on wheels I had ever seen on the streets or in photos. Seeing a strange vehicle in Detroit was not unusual, and I assumed this aerodynamically shaped vehicle was just one of the many prototype cars often appearing on the streets.

The body contour flowed rearward fitting within the running gear to enclose the rear axle. It surrounded the rear wheels and included that lovely dorsal fin centered on the top rear. The front wheel covers, which pivoted to allow clearance for wheel protrusion on turns, indicated an exquisite attention to detail rarely found. It was quite different, yet at the same time, it appeared to be very logical.

The car had no nameplate to identify it (a common practice back then). I assumed the owner was here to present his new design to the numerous candidate companies in the Motor City. Years later I learned what I had seen was a French Dubonnet "Dolphin," the only one of its breed, brought to the United States to demonstrate to Henry Ford. Neither Ford nor General Motors were interested, and the little Dolphin sank into obscurity. (Appendix I, A Brief History, contains a more detailed description and photo of the Dolphin).

There were many also-rans and prototypes on the streets of Detroit in the 1930's. The auto industry was very extroverted in those days. A logical tendency toward secrecy was countered by an exuberance which said, "just look at this!" In contrast to today, there were countless car makers all vying for attention. Even then "Detroit" was a generic term for the U.S. auto industry. The Nash was made in Kenosha, Wisconsin, the Studebaker in South Bend, Indiana, and the Willys Overland in Toledo, Ohio. Still, Detroit was magnet, a focal point of industry activity.

Experimental engineering concepts were tested out on the streets. I can recall seeing the custom roadster of Edsel Ford's, based on a 1933 Ford V-8, tooling down Jefferson Avenue. Another time, a Willys (the make made famous by the WWII Jeep) sedan was parked outside my school. A close inspection revealed independent suspension on all four wheels tucked in underneath an otherwise unassuming body. A relatively advanced concept for the 1930's, the advantages in riding and handling offered by that suspension system might have given Willys an invaluable edge in the market place if it had been put into production.

Not all the cars I saw first revealed on my neighborhood streets were failures, of course. In the summer of 1938, I was startled by the sight of the first Lincoln Continental, custom built for Edsel Ford on a Lincoln Zephyr chassis, sitting on Lake Shore Drive. In terms of pure style, this low-profile craft, with large windows, thin roof line, narrow pillars and voluptuous contours was an inspiration. I learned later that Edsel Ford had commissioned the design from E.T. Gregorie for use by the Ford family in Florida. This *one-off* was such a resounding success that its design was put into production and became a classic. Some people regard the Lincoln Continental as the loveliest automobile design ever (photo 53, Appendix I).

The period from 1934-38 was, in my opinion, the golden age of Detroit automotive genius. That period saw the end of the carriage era in automobile bodies with the disappearance of wood from frame structures, and the end of composite roof panels. "Steel alone is not enough!" was the typical claim by some automobile manufacturers, who used wood to re-enforce frame tubing and body sections well into the 1930's. Structural and manufacturing advances made the all-steel, smooth welded steel roof commonplace. Solid front axles began to vanish in favor of independent front wheel suspension, yielding significant improvement in riding comfort and handling. The two-piece "V" shaped windshield replaced the flat front windshield, and glass area increased as pillars thinned. The radiator retreated into the hood behind a grille, and the luggage trunk blended into the body. Aircraft-style instruments (illuminated from within) came into vogue, while interiors emulated contemporary trends in furniture. All these improvements developed at the behest of a new breed of professional: *the Automotive Stylist.*

Some engineers took pity on the motorist struggling to see through steamed-over windshields in rain and cold by integrating defroster ducts into the interior garnish molding. Others ended the visage of bundled, shivering occupants by inventing forced-air heating systems which made the interior bearable in severe weather. One of these heating systems was the Nash *Weather Eye* first demonstrated in the late 30's. I can recall seeing these Nash automobile demonstrators cruising downtown New York and Detroit streets in mid-winter. The right rear door had been replaced by a clear plastic panel. Passersby could see a lovely model clad in summer garments seated inside, beaming with joy in the comfort of a Nash Weather Eye forced-air environment — a first in automobiles. The snow may have been flying, but pedestrians, huddled against the elements, could not help but see the balminess inside that car.

This was also a period of great economic travail. The consequences of the stock market crash of 1929 continued to weigh upon almost every shoulder. The automobile industry, stunned by plummeting sales in the early 30's, recovered, and with 1933 and 1934 models, stimulated sales with the new ideas and new designs of the Golden Era. There is no doubt that this courageous stand against adversity played an important part in revitalizing all United States industry from 1934 onward.

However cheerful and innovative the Golden Age of Detroit Design appeared to be, the progress actually being made only emphasized how much more could be done. Those years of close acquaintance with the vitality of Detroit's automotive progress, thoroughly inspired me to be a part of that exciting drama some day, somehow. There were many engineers and designers who were shouting "more!" and I wanted to be among them.

Automobiles were not to be my first brush with design, however. My father's career brought us back to New York City in 1937. I attended McBurney High School in Manhattan, enjoying its stimulating atmosphere. My father died in 1939 just as I was about to graduate. I was on my own at age eighteen.

Thanks to my father's insurance benefits, I was able to stay in New York and plan college level studies. Aircraft design beckoned — a love parallel to that of automobiles. I enrolled in Stewart Technical Institute on Manhattan's west side. The curriculum was excellent, the instructors top notch, but stress analysis of structural joints was a crashing bore. I wanted to design whole objects, not minute parts. After nearly a year of what was actually invaluable learning at Stewart, I left to pound the streets of New York in search of industrial design work — that then-new facet of industrial art which sought better answers to how objects should look and perform in human hands. After countless interviews, a fortunate reference by my father's best friend, Jack Lucas, led to a job.

Harry Preble, Jr., a young architect from Denver, Colorado, had begun practice in architecture and product design in a tiny office on West 42nd Street in New York City. I was hired by Mr. Preble in late 1940 to help him manage a temporary work overload. This work included a speculative commission from General Electric for a pocket radio, as well as a confirmed project for the A.C. Gilbert Company (a then-famous scientific toy company) for the total redesign of their headquarters — the former Outdoor Advertising building on lower Fifth Avenue.

These and other projects were fascinating for a thoroughly neophyte designer. Preble was a most charming talent, an excellent teacher and superb renderer in pastels. He introduced me to many remarkable and innovative architects and designers in New York in 1941.

I enjoyed only a brief start in the field of design before the United States entered into World War II. The pressures on the American economy for Allied armaments, the obvious trend toward U.S. participation in the European war, plus the ominous threat of war in the Pacific, were steadily mounting. The growing war effort diminished

Harry Preble's need for an assistant.

The General Electric project was one of the early casualties. Some years later others would claim that they had invented the miniature portable radio. Actually, it was only the onset of World War II that prevented General Electric from finishing their pocket radio project. Our designs for a GE portable — based on GE's working prototypes — were about the size of a 1988 belt-worn paging device. The small battery-powered portable radio — in somewhat larger form — was commonplace in this country before the war.

Reluctantly, I left Harry Preble Industrial Design, and after a brief stint in advertising art, went to work for Edo Aircraft on Long Island as an engineering designer. As my formal schooling was in aeronautical engineering, I was on solid ground. At Edo, I was involved in the design of amphibious floats for the Douglas C-47 (the military version of the already famous DC-3). Edo was an inspired company, headed by Edgar D. Osborne, the leader in the design and manufacturing of aircraft floats. The firm was abundantly staffed by Russian emigres who had departed their homeland during the 1917 revolution. Their chief engineer, B.V. Korvin-Kroukovsky, was a consummate engineer, who maintained a standard open-door policy for subordinates, encouraging them to ask questions and vent their views. I learned a great deal from taking advantage of that policy.

Motivated by the now rampant War Fever, the urge to fly in the Air Corps was overwhelming. Within one year, I went into the service to learn how to be a military pilot. Upon my leaving Edo, Korvin-Kroukovsky presented me with a letter of recommendation which I treasure in my collection of mementos.

During my three years in the Air Corps (March 1943-March 1946), my enthusiasm for aircraft design subdued my other passion — automotive design. However, two protracted periods of training in piloting and electronics allowed me ample time to dream of cars, especially of an ideal post-war vehicle. Conditions in the service were not very suitable for leisurely sketching. I did not have a large, well-lighted drafting board at my disposal, but the hurry-up-and-wait syndrome of military life left much opportunity for thinking, and a "borrowed" breadboard became my design studio. Every otherwise "wasted" minute, spent standing in formation, drilling, or waiting for something to happen, was

devoted to working out the details of what I christened *"Excalibur."*
My ideas, sketched out on any available paper and drawn up on my
breadboard, became my dreamed-of "Master's Thesis" — a book of
about fifty pencil renderings and one large water-color.

I added my own ideas to the list of what I considered to be exciting
contemporary automotive designs. Excalibur would be a small car. Its
rear engine would be mounted transversely directly over the rear axle
centerline, with a drive gear at the center of a six cylinder crankshaft,
coupled to a transmission below. Other carefully planned features
included independent suspension on all four wheels, a turbo super-
charger, retracting headlights, a sun roof, and the elimination of the
dash board below the crash pad on the right side. In styling terms, the
belt line (bottom edge of the window openings along the sides), was so
low that, when rolled down, the window glass would touch bottom at
the lower inside of the doors.

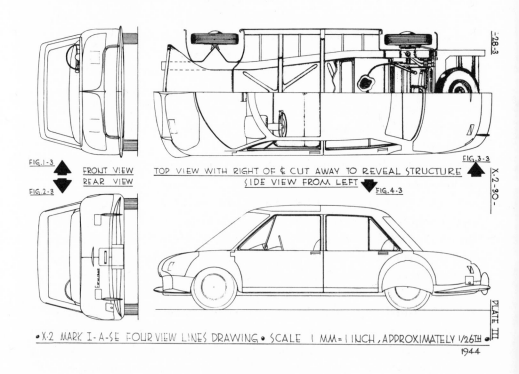

FIG.1-3   FRONT VIEW   TOP VIEW WITH RIGHT OF ℄ CUT AWAY TO REVEAL STRUCTURE   FIG.3-3

REAR VIEW   SIDE VIEW FROM LEFT   FIG.4-3

FIG.2-3

• X·2 MARK I·A·SE FOUR VIEW LINES DRAWING • SCALE 1 MM = 1 INCH , APPROXIMATELY 1/26TH •

1944

"Excalibur"

The exercise was splendid self discipline. I learned a great deal about automobiles from concurrent research, and much about my fellow human beings from living in the typical tight quarters of the service. Rather than regarding me as the odd-ball of the squadron who never went into town on his free time, but who instead stood on his foot locker with his breadboard at a comfortable height on the upper bunk, my mates were immensely curious and complimentary. It was interesting that a number of those who were ordinarily regarded as tough guys with nary a good word to say about anything, were among my best friends and allies.

The end of World War II made the realization of countless dreams seem possible once again. Reunion with loved ones could be realized. Return to home, to office, factory, and farm ceased to be a fantasy dreamed of between battles, and became a reality for those fortunate survivors of the intense trauma of the war. For the victors, such aspirations were now possible. For the vanquished, it would be many years before dreams of normalcy even made sense.

In the United States, the cessation of hostilities meant an overnight cancellation of virtually all government contracts for war materials. To prepare post-war industry for the wondrous prospects so heralded in the press, shortages of materials and parts had to be overcome, manufacturing facilities reconverted, product designs created and markets reopened. Most companies swiftly resumed pre-war levels of production.

World War II, the bloodiest conflict in history, was also the crucible for an incredible number of technological advances. It provided the impetus for an ever increasing tempo of development, like a motion picture being run at double time. Accomplishments in productivity and the successful application of science to engineering practices made the future appear like the sun coming up on a vast cloudless horizon. Rocket propulsion, turbojet engines, atomic energy, radar, hydraulic drives, and a myriad of other developments had progressed from theoretical concepts to reality during those years. With them came innumerable potential benefits to mankind in transportation, metallurgy, electronics and mechanisms. What prevailed at the end of World War II was not just reconversion, but the enthusiastic embracing of new and better products.

The automobile industry's dreams and designs were postponed by the war. In 1945, the dreamers once again emerged on the scene, some with loyalty to older concepts. These included engines in the rear of the automobile (which had never fully gained recognition) and even steam power for cars (which had failed utterly in the past).

The popular media continued its enthusiastic portrayal of automotive things to come, promoting every conceivable form of vehicle as the wave of the future. There were three wheel cars, front wheel drive cars, air cars, electric cars, even atomic cars. Many of these ideas were not new, a number were quite demonstrably valid — and some ideas were blue sky.

Even though I knew my ideas were valid, Excalibur was stillborn in terms of tangible fruition. Even its name was post-empted by a famous automobile designer, Brooks Stevens, who no doubt, was totally unaware of my dream car. Ford, General Motors, and the others to whom I wrote about my project never responded. The inner knowledge I felt in having designed *the* car of the new era was my only reward. (photo no. 1)

CHAPTER 2

ENTER PRESTON TUCKER

Among the many dubious new automotive concepts publicized by the press were some *apparently* valid ones. During the war, hydraulic drives had been developed for gun turrets, such as those on tanks. The application of hydraulic power to moving mechanisms reached a new sophistication. There were many advantages to hydraulic drives in comparison to mechanical drives: quietness and the cushioning effect of fluid drive from a remote source of hydraulic pressure connected by hoses with intervening controls (such as forward, stop, reverse) to a compact remote motor. Automotive applications, such as hydraulic power steering, were thought to be endless.

A cursory awareness of hydraulics can lead to some glowing conclusions. These began appearing on the pages of many popular mechanical and scientific magazines. Inventors' dreams of hydraulic drive for automotive vehicles promoted the idea that a large engine-driven pump could send hydraulic fluid, under great pressure, through hoses to hydraulic motors in each of two or four wheels of a car or truck propelling it with smooth flowing power. There would be no need for a transmission, prop shaft, differential, and driving axles; and the resulting simplicity lowers the cost.

One of the great believers in the theory of Flowing Power was Preston Thomas Tucker, an engineer who for many years had been a friend and associate of racing car designer, Harry Miller. Tucker started his career in the automobile game as an office boy at the Cadillac Division

of General Motors in Detroit. His devotion to cars led to the racetrack
where he became a confidanté of Miller and a respected friend of the
Miller family. A family man, nearly forty years old, Preston Tucker did
not see active duty during World War II. However, he was active with
the Washington D.C. military staff and designed mechanical projects,
including work on the Sherman tank, his radically speedy combat car,
and his famous "Tucker Turret" which became a standard in the war.

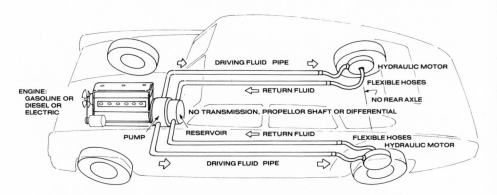

**WHEEL SUSPENSION AND POWERPLANT
DETAIL OMITTED FOR CLARITY**

Flowing Power was the idealized simplification of getting power to the drive
wheels, and only worked in theory.

Tucker's apparently detailed understanding of fluid power underlay
his concept of his post-war car. In 1946 he began to promote, in press
releases and brochures, the idea of an engine-driven pump which
would deliver hydraulic fluid to hydraulic motors at the rear wheels to
propel a rear-engine automobile.

At the end of the war, Tucker was one of the visionaries with a well
developed dream. He didn't simply want to sell an idea (such as fluid
power) or an invention or two to existing automobile manufacturing
companies. He wanted to create a company of his own. A company in
which he could design, engineer, produce and market a car which
would revolutionize the entire automobile industry.

Preston Tucker was no ordinary visionary. He had lived his life in the midst of cars, all cars, but especially race cars. The scent of the Indianapolis Motor Speedway was part of his presence. His association with Harry Miller was akin to being an aeronautical acolyte to Charles Lindberg. Miller's credentials included eleven Indianapolis Speedway wins in the 1920's and 30's. His were the apotheosis of high-speed vehicles capable of enduring and excelling in a race which, in a few hours, demanded a trip at maximum speed, equivalent to half the distance from Chicago to New York without a significant hitch and with few stops.

From this association, Tucker acquired an impressive store-house of information. He felt that he could now apply this knowledge to ordinary cars. He knew what was not quite right about the automobiles then traveling the highways of the world. In Tucker's view, the ordinary car was too slow, too top-heavy, its suspension inadequate. It was dangerous to occupants in an accident, its engine was not located ideally, its brakes were inefficient, its engine was excessively fuel-hungry, and its shape did not smoothly cleave the air through which it ran.

Most of these characteristics were well known to others of course. Tucker was following the lead of profoundly professional talent such as Jaray, Ledwinka (designers of the Czechoslovakian Tatra), and Porsche, when he set out to produce his wonder car.

He was also venturing into a lethal arena. Before Tucker, the revolutionary Tatras had failed to revolutionize. Sleek Silver Arrows had quickly shot across the pages of automotive history, lovely Cords had come and gone twice, and aerodynamic Airflows had not captured the acceptance of the buying public (see Appendix I for photos, descriptions of these automobiles).

Tucker faced a challenge that demanded every aspect of his experience, his talent and his business acumen. In a manner which seemed unique to him he claimed to be able to see through most of the problems of past automotive designs, and to have the answers to assure his success.

His name began to appear in magazines and newspapers with increasing frequency during the immediate post-war period. Press releases attributed to him repeatedly heralded new ideas of fluid drive, rear engine location, and novel styling concepts. A host of features were soon to materialize on the new Tucker production lines. Tucker's entreaties were directed toward an avid audience. There were countless enthusiastic would-be car dealers and customers who wanted to sell and buy a post-war marvel.

Passenger car production lines in the U.S. had ground to a halt in February 1942, and resumed in July 1945 with essentially warmed-over early 1940's designs; a purely stop-gap measure to get vehicles to market quickly. This was a tantalizing delay to the world-wide appetite for refreshing change, but it shouldn't have been a surprise.

Not all of the activity within the walls of major automobile company buildings was devoted to the war effort from 1942 to 1945. Small staff groups did significant work and study on post-war automotive designs. Much of it was devoted to cosmetic surgery on the 1941 models. Numerous clay studies were executed with an eye to getting the plants back into production without the delays required to manufacture new expensive forming-and-stamping dies for an entire new body. A new grille perhaps, new colors and trim at first. New designs could wait. What people would want would be cars — transportation — period, paragraph.

Thus, Preston Tucker's timing seemed to be perfect. What better opportunity could there be for a revolutionary to get the jump on an industry which obviously was not going to rush into the production of new designs in spite of the great technological advances during the great war?

CHAPTER 3

REBIRTH

Having spent six months in Europe after the end of hostilities, I had only a vague idea of what had been going on in the world of design during my absence. Most large design firms had managed to adapt to wartime conditions. Norman Bel Geddes, whose fame had spread from stage design to proposals for revolutionary ocean liners, had survived the war by creating splendid dioramas. These fine miniature sets of World War II great sea battles were given prominent two and three page coverage in *Life Magazine*. This was typical of the enterprising approach necessary to see the war years through. Many design outfits made instructional films, undertook human engineering studies and introduced the problem-solving abilities of industrial design as a profession to the military.

Re-entry into the design world of 1946 New York was a challenge. Not long after returning to the states, I followed through on a contact at J. Gordon Lippincott and Company at 500 Fifth Avenue, New York. A week of so earlier, I had met Gordon Lippincott in his lovely corner office on the fifty-fifth floor, where I showed him my rudimentary portfolio, put together in countless midnight oil efforts. He had not turned me away, but alluded to the numerous active projects which his young design firm had underway —and he hinted at possible employment. One of the active projects was for Andrew Higgins (the WWII torpedo boat magnate, who wanted to go into the pleasure craft business). He and Gordon Lippincott were in frequent contact with each other.

A week later I made a phone call to follow up on this possible job offer. I asked the receptionist if I could speak to Mr. Lippincott.

"Who is calling?" she asked.

"Mr. Egan", I replied.

I was immediately put through to Gordon Lippincott.

"Yes, Andy?" came his clear voice.

"Umm, Mr. Lippincott, this is Phil Egan. I came in to see you a week or so ago about a position with your firm."

Gordon, a gentleman always, recovered in an instant upon the realization that he was not talking to his friend, Andrew Higgins. Little did I know, at that time, that Lippincott had a client whose name could be confused with mine by a telephone receptionist. Later, I also discovered that Andrew Higgins was a close friend and former business associate of Preston Tucker.

I recovered in kind to press further upon the possibility of employment. Gordon suggested that I come in again to see Walter Margulies, his associate, and Read Viemeister, his director of styling. In the meeting that followed, I met Mr. Margulies, a dapper, cosmopolitan gentleman with a charming European accent, who studiously viewed my portfolio of largely speculative works. He called in Read Viemeister, a handsome young man, not quite my age, whose long sideburns were an immediate distinguishing feature, to appraise my efforts. I recall that Read's comments about my renderings were that they were excellent but "tight". "Tight" meant time-consuming in those days, when extraneous fabrication considerations were not a part of a presentation to a client. Loose pastel sketches, if done with a masterful hand, such as Read's, sufficed in presenting ideas. Renderings — the presentation of a design idea in art form — are usually finely drawn and more elaborate than sketches.

Nonetheless, Margulies hired me, and, on a Monday in June, at age twenty-five, I went to work for J. Gordon Lippincott and Company for the magnificent sum of seventy-five dollars a week. This was exactly twice my income at Edo Aircraft. Times *had* changed in three and one-half years.

Working at Lippincott meant many diverse design projects, from model making of automatic turntables, to sculpting fountain pens on a lathe, to mocking-up truck trailer housings. It also meant meeting and working with a talented group of individuals. Gordon Lippincott had been a professor of design at the Pratt Institute in Brooklyn, just over the East River from his new location in Manhattan.

Pratt was one of the pioneering industrial design schools in the United States and it produced an impressive roster of graduates who went into their chosen profession with notable success. Among those were members of the Lippincott staff — Read Viemeister and Budd Steinhilber — who joined Gordon immediately after their graduations in 1943 and quickly rose to important positions in the firm. By 1946, Read was Director of Styling and Budd his closest fellow design associate. Read had progressed from graduate to director in three years through an enviable spectrum of design projects: aircraft exterior color schemes and interior designs, automobiles, and the Nautilus submarine interiors. Lippincott was commissioned to do the living quarters design of the first nuclear submarine ever built. Budd had shared in some of these efforts, though his work was interrupted by wartime service, and he had only recently returned to the firm. At the time I joined Lippincott, Budd was active in many creative projects. I recall his executing some of the finest Prismacolor (colored pencil) renderings I had ever seen. They were of entertainment foyer concessions, and they seemed to dance right off the black mattboard upon which they were drawn.

Then there was Tucker P. Madawick, a Pratt student who went to the Ford Motor Company as an apprentice designer in 1938, and became part of the Mercury car project. The Mercury was Ford's first additional automobile line (the Lincoln having been an acquisition). At the onset of WWII, Madawick became enmeshed in the wartime activities of Ford, including the production of Consolidated B-24 bombers at the mammoth Willow Run, Michigan plant. He subsequently joined Consolidated itself in Fort Worth, Texas. There, he was active in the production of B-24s, the B-32 alternative to the famous Boeing B-29 Super Fortress, and the monster B-36 bomber (which prevailed into the 1950s). The moment World War II ended, Madawick went back to New York to join his former design teacher at Pratt, J. Gordon Lippincott, in the new aerie in Manhattan.

Tucker Madawick was assigned to the Republic Aviation account. Republic (formerly Seversky) had become famous as the producer of the Jug, the legendary P-47 fighter, designed by Alexander Kartvelli. In World War II this plane met and matched the best that the German Luftwaffe sent into the skies over Europe. Republic's post-war sights were set on two aircraft. One was the SeaBee Amphibian, a single-engine, four-seat private plane powered by a single Franklin air-cooled engine. This plane was expected to capture a vast market. Lippincott designed the interior of the craft. The other was called the Rainbow, a four-engine 400 mph (640 km/h) transport. Originally built as the XF-12, a high altitude photo reconnaissance plane for the Air Force, Republic was now grooming it to be the spearhead civilian airliner for the world's commercial airlines. Tucker Madawick and other Lippincott staff members worked on converting the interior design of the plane to comply with civilian tastes.

Another member of the Lippincott design roster was Hal Bergstrom, a seasoned veteran of industrial design, fresh from Raymond Loewy Associates (Loewy's name was substantially the best known in industrial design at that time). Hal was the cool, calm, collected type who took an assignment in stride and always came out on top. These attributes would be put to the test in a near future automotive assignment. When I joined the company, Hal was in charge of the Northwest Airlines (later, Northwest Orient) account, which involved exterior color schemes and the interior design of the passenger sections. He was also active in the Republic Rainbow project. In those immediate post-war days, aircraft interior design soared as high as the planes, with comfort and luxury the passwords.

Lippincott had a completely-equipped model shop with staff at another location in west side Manhattan. Besides those men I came to know well, there were several other designers, two expert mechanical engineers, and two highly qualified heads of the organization, Gordon Lippincott and Walter Margulies. Walter had come to the firm with many credits in interior and architectural design. He had a keen awareness of trends in design and was highly adept in dealing with all of the nuances of client relations.

I quickly learned that the Lippincott group wasn't just conversant with record turntables, fountain pens and truck trailers. In the closing days of the war the company made several auto industry contacts. They had already worked on a post-war car. Joseph W. Frazer, the head of Graham-Paige in Detroit, hoped to revitalize what had become a moribund automotive enterprise in the 1930's. He wanted to be the first with an advanced post-war vehicle. Lippincott secured the prospective design contract, and gave Read Viemeister the assignment.

The initial presentation to Frazer was a *full-sized* rendering done in New York. Lacking space for such a grandiose undertaking in the 55th floor studio, Read and his assistant Jay Doblin executed the creation in color pastels on the floor of the elevator lobby after business hours. Read said that his and Jay's fingers were rubbed practically to the bone by the exercise (pastel renderings are usually done by applying the color directly onto paper via the pastel stick and then smoothing and blending the medium with fingers). Read then went to Detroit to develop numerous smaller renderings and a one-quarter size clay model in the Graham-Paige plant. I saw the 8x10 photos of the clay model only a few days after joining the firm, and it was clear that there was genuine talent in the Lippincott ranks. There were many exciting ideas in those renderings and the clay model. (photo no. 2).

Even though Read's designs did not go directly into a Graham-Paige car, I could see Read's influence in the final results. Joseph W. Frazer of Graham-Paige teamed with Henry J. Kaiser in 1945 to build Kaiser-Frazer automobiles in the former B-24 bomber plant at Willow Run, Michigan. Howard Darrin was the ultimate designer of the car, and accomplished the task with dispatch.

Read's associate, Jay Doblin (also a Pratt graduate) went on to an illustrious career in design at Raymond Loewy Associates, as the director of the Institute of Design of the Illinois Institute of Technology with Mies Van Der Rohe, and later as an independent consultant.

By a great stroke of luck, I had landed in the midst of a New York design outfit which was not only deeply involved with one of my loves, automobiles, but with another, aircraft. I couldn't have made a better arrangement on purpose.

CHAPTER 4

THE EMERGENCE OF TUCKER

"TORPEDO ON WHEELS" shouted the headline in the December 1946 issue of *"Science Illustrated."* The article went on to claim, "Engine in rear, all-hydraulic drive make the Tucker a real car of the future." A striking rendering of a dramatically styled coupe fairly leaped off the page and into the reader's future. The car, featuring an aircraft-style cockpit with apparently fixed side windows, a center head-light, and all-enveloping fenders mounted so they pivoted with the wheels, could only belong to some far-distant time and place. Yet there, in the upper corner of the page, was a photo of what appeared to be the actual car or, at least, a full sized mock-up. The copy prom-ised that "The Tucker Torpedo is scheduled to hit the road sometime in '47... This startling car will incorporate spectacular engineering innovations that conservative auto manufacturers have classified as at least five or six years off." As partial proof of this claim, in the lower corner of the page was a drawing of Preston Tucker's hydraulic drive with hoses running from an engine-driven pump to hydraulic motors in each of two wheels. (Photo no. 3)

The advertisement revealed that this was no ordinary car, certainly not a warmed-over pre-war model. The photo — a three-quarter rear view—showed a decidedly aerodynamic automobile. The car emphas-ized art nouveau speed lines and shapes. The article did not mention the designer of the car, George Lawson.

Science Illustrated was no sensationalist tabloid. A certain degree of credence was implied to any new idea or product mentioned on its pages. For any automobile enthusiast, this was heady stuff. Was Tucker really on the way? Did he actually have a car being crafted in steel, with a developed engine, and the funds to bring it into being, and the facilities and staff financed by these funds? In the 500 word article, Preston Tucker seemed to state categorically that he was, indeed, about to carry off his challenge to the automotive industry. This publicity was widely disseminated in 1946 periodicals. Press releases, advertisements and brochures all promoted the revolutionary Tucker Torpedo.

Not much more was heard in the national media from the Tucker works until, March 2, 1947. Full page advertisements in *The New York Times*, *The New York Herald Tribune*, and many other newspapers, heralded what seemed to be significant Tucker progress (photo no. 4).

The ad was no longer for a "Torpedo" but simply the "Tucker '48." It was somewhat different from the "Torpedo" of a few months before, a bit more conservative. The rendering showed a vehicle with fixed fenders, four doors, and orthodox side windows. The cycle fenders were no more, but the center headlight still remained. The hood was sharply tapered with a dramatic lower grille. The body retained the severely tapered back of the Torpedo. The prominent rear fenders were adorned with trim. Both front and rear fenders had a novel feature, an upward jog in shape just aft of the wheels, almost like the signature of a designer. Again, there was no mention of the designer of the car. This time it was Alex S. Tremulis.

The headlines and text of the advertisement were assertive and convincing: "How 15 Years of Testing Produced The Surprise Car of the Year. You'll be smart if you wait. . .Later in 1947, you'll judge the new car for yourself." The impressive ad claimed that pilot models had features completely proven by fifteen years of rigid tests, including, "FLOWING POWER sure as a mighty stream, moving directly from engine to wheels, as free as a seagull's glide." The Tucker was to have single disc aluminum brakes, individual wheel suspension, fuel injection, precision balance with hairline steering and driver control. Unique safety features were promised including a shatter-proof windshield that popped out on contact, foam-padded interior with crash-padded dashboard and a *safety chamber* in front of the front seat passenger.

The ad also claimed the Tucker '48s were being completed in the largest, most modern plant in the world. This last claim Preston Tucker could make with confidence. He *had* acquired a home for his enterprise — the vast Dodge plant on South Cicero Avenue in Chicago, Illinois. This plant, then the largest single building complex under one roof, was built to manufacture Wright double-row Cyclone engines used in B-29 bombers. After the war, the plant was placed in the hands of the War Assets Administration — the government agency entrusted with all property, from drafting tables to entire plants, built by the United States government during the war. Tucker had signed a lease with the WAA in July, 1946 for the 475 acre plant, contingent upon his being able to raise $15,000,000 in capital by March, 1947. Tucker quickly set in motion an ingenious program of raising capital through a dealer franchise program plus a stock issue of $20,000,000 through the Chicago brokerage firm of Floyd D. Cerf. Entanglements and counter-claims on the property delayed the formal acquisition until September of 1947, even though Tucker had succeeded in raising the capital.

Within a matter of months, two generations of Tucker automobiles had been presented to the public as *fait accompli.* The differences in the two advertised Tucker designs were pivotal. The men responsible for the two generations of the car — which didn't yet exist — were both highly qualified automotive designers. As I looked over the two different advertisements, I suspected the divergence of opinion which the two designs represented indicated that Preston Tucker might still be in the process of deciding what the Tucker automobile would finally look like. This proved to be a valid surmise.

George Lawson's credentials were those of a genuinely talented artist. Schooled in his profession at the Cleveland School of Art, with a portfolio of experience in commercial art, illustration, and automobile design at General Motors, Lawson made notable contributions to sketching and rendering techniques which were adopted by many other designers. He was in charge of design for the Buick division of General Motors in the late 30s. During that time Buick went from pedestrian to innovative. Lawson later found a niche in the design studio of the Briggs Manufacturing Company. It is not commonly known that Briggs (and other firms, such as the Murray Corporation of America) designed and made automobile bodies for name automobile manufacturers, who put them on their own chassis or sometimes simply installed engine and running

gear and marketed the product as their own. An interesting twist to
Lawson's part in the Tucker story is that at Briggs, he met Alex Tremulis, also on the design staff there, although this connection had nothing
to do with their later, separate roles in the Tucker story. During WWII,
Lawson was back at GM, but not as a stylist. Instead, he was chief of a
section designing and producing training manuals for the government.

In 1945, Preston Tucker was often in Detroit seeking backing to help
create his company to produce his post-war dream car. Tucker needed
tangible evidence of an actual design to excite potential backers and to
justify venture capital. He heard about George Lawson, contacted him,
looked over his extensive portfolio and commissioned him to design
what became the Tucker Torpedo.

Lawson carried the project through innumerable sketches and renderings, plus a one-quarter scale clay model, which was carefully painted
and then photographed against outdoor backgrounds to simulate reality.
The photos were skillfully retouched, resulting in convincing realism.
This was *the car* seen in press releases heralding the Tucker Torpedo,
which would be "in production in 1947." (photo no. 5)

The merits of Lawson's design have been debated ever since. It was a
dramatic *tour de force*, calling for cycle fenders at the front, each with a
headlight which would illuminate a dark sector in a turn. This was a
seemingly logical extension of a popular accessory of earlier expensive
cars — pivoting headlights, mounted above bumper level, that turned in
the direction of the front wheels, connected to the steering wheel by
linkage. They were an attractive embellishment for a Packard Phaeton or
a Stutz Black Hawk. Cycle fenders were not a new idea either. They had
long been used on sportcars with distinctly separated fender and hood
shapes. Lawson was proposing an integrated, flowing shape with everything filled in. It looked good on paper, but it was never explained how
the forms would allow the fenders to telescope in and out, regardless of
dirt, gravel, mud or ice.

Another feature of Lawson's Torpedo was the position of the driver.
He or she would be in center front — no more eye-balling the center
divider on the road. Viewed from above, the vehicle looked like a
double-ended rowboat with turning front fenders and rear fenders gracefully blending into the sharply tapered rear.

The projected specifications of the earliest Tucker rendition were:

Wheelbase: 126" (320 cm)
Front wheel tread: 64" (163 cm)
Rear wheel tread: 65" (165 cm)
Horsepower: 150
Engine placement: rear
Driven wheels: rear
Engine weight: 250 pounds (113 kg)
Car weight: 2000-2400 pounds (900-1100 kg)
Cruising speed: 100 mph (160 kph)
Top speed: 130 mph (209 kph)
Fuel economy: 30-35 mpg (15 km/l)
Projected price: $1500. U.S.

No mention was ever made of the consequences of Lawson's sharply curved side windows with respect to side access to pay a bridge toll or just pass the time of day with a friendly highway patrolman who sought a word or two about the 100 mph cruising speed. How the windows would retract to provide such access was never explained. In recent years, this same problem confronted another would-be auto magnate, John DeLorean. His solution to the fixed side windows was a small access port in his gull-wing doors through which one could, at best, pass money to a toll-taker. This might have been the same solution for Lawson had his design reached production levels.

There does not appear to be any record of design discussions between Preston Tucker and George Lawson. It can be noted that Lawson was forced to sue Tucker for payment for his work, and he recovered a portion of his fee. However, Tucker was able to use the fruits of George Lawson's sincere efforts in initial press releases and brochures showing a tangible car, which set the stage for dealer franchise deals and a stock float. The essence of Lawson's design remained for further development by another designer, one Alex Sarantos Tremulis.

Alex was an equally talented designer, with credentials going back to the legendary Duesenberg. In 1933, at the age of nineteen(!), he joined Duesenberg in their Chicago office as a stylist, where he assisted in adapting custom-made Duesenbergs to the specialized tastes of its often exotic customers. By 1936, he was their chief stylist at the manufacturing plant in Connersville, Indiana. He earned his stripes at that company

(one of the Auburn-Cord-Duesenberg trio) by his adaptations of existing Duesenberg and supercharged Cord 812 designs to exposed exhaust
pipes — gleaming flexible tubes which exited from the hood side and
connected to pipes going to the rear of the vehicle.

Duesenberg was an anomaly in the history of American automobiles,
one that miraculously survived the depths of the Great Depression of
the 30's, only to perish just as economic conditions started to improve
under President Franklin D. Roosevelt. The gallant car was a kind of
U.S. Mercedes Benz, with features, quality and performance unexcelled
at the time. Beginning in 1920, Fred and August Duesenberg produced
high performance automobiles with a luxury price tag which were to
become unique in the United States. E.L. Cord, the famed entrepreneur
of Auburn and Cord automobiles, took over Duesenberg in 1926 and
made the "Duesie" the prima car of the next decade. Alex Tremulis has
said that the 1934 Duesenberg SJ could do over 100 mph, (160 kph) in
second gear when most of the passenger cars in the world could not
make 90 mph (144 kph) in high gear. It became the car of celebrities,
especially Hollywood celebrities. This distinction was probably the
secret of its survival through the Depression. These few people, motion
picture stars among them, made enough money in hard times to be able
to afford expensive cars and the Duesie was one of their pets. The allure
was apparently just a gloss, and when the economy began to improve in
the late 30's, Duesenberg could no longer maintain its charisma. In
1937, the company closed its doors forever, even though others have
attempted to revive the name in recent years.

After Auburn-Cord-Duesenberg shut down production, Alex Tremulis went to General Motors in their Oldsmobile division and then to
Briggs Manufacturing, his first ventures into the Motor City. He was an
innovator with pencil, pen, and his skill with that tiny cousin of the
spray gun, the airbrush, is legend. Others used masking frisket paper to
apply successive layers of color and shape; he used a triangle or a French
curve to guide the spray and produce magnificent renderings in a frac-
tion of the usual time.

In 1938, Tremulis joined Custom Motors of Beverly Hills, California
(owned by Eleanor Powell, the famous dancer) as a stylist. For two years
he learned first hand of the numerous peccadilloes of Hollywood celebrities who sometimes had more money than taste.

In 1939, Alex joined the American Bantam Car Company in Butler, Pennsylvania as a consultant stylist. The American Bantam was an adaptation of the English Austin, a miniature car far ahead of its time. The English version of this vehicle had been introduced into the U.S. market virtually unchanged and had not been successful, even though it was featured in a famous 30's Hollywood film, "*A Connecticut Yankee in King Arthur's Court*". The film, starring Will Rogers, was a hit; the car wasn't. Under new ownership and name, the car was restyled by Alexis de Sakhnoffsky. Alex Tremulis designed a two door convertible version, the Riviera, for Roy S. Evans, the new owner. It was so appealing that it was placed in production. The American Bantam was a valid, fully roadable small car, years ahead of its time.

During World War II, Tremulis was the chief illustrator of the Aircraft Illustration division of the Aircraft Laboratory at Wright Field, Dayton, Ohio. Here, from 1941 to 1945, the young master sergeant was privy to most of the early developments of rocketry and jet propulsion. Wright Field duty for a designer in the military was like going to heaven without the bother of dying. While there, Alex recalls, ". . .I was responsible for the first two-stage vertical launch vehicle, code named TVT (for Tremulis Vertical Take-off). Later it became known as *Operation Dyna-Soar.*" The Dyna-Soar concept evolved into a funded project that progressed to the hardware stage as the hypersonic boost glide program under the management of the Boeing Airplane Company before it was dropped — temporarily. The Dyna-Soar was the direct antecedent of the Space Shuttle, differing fundamentally only in size and number of crew. (photo no. 6).

Alex Tremulis' aerospace work was so invaluable that he was re-employed in early 1946 by the Air Force as a civilian for the exciting birth of post-war aeronautics. When severe budget cuts ended his civilian duty with the Air Force, Alex returned to Chicago and joined the industrial design firm of Tammen and Denison — a move which was to have a profound influence upon his career and upon automotive history.

At about this time, Preston Tucker had picked up his proceeds from the George Lawson efforts, established a stake at the huge Dodge aircraft engine factory in Chicago, and was ready to get down to a closer relationship with reality. It was late 1946 and time was pressing hard on

Tucker. The post-war vacuum of automobile production was being filled with ever-increasing vigor. Production of warmed-over pre-war designs began to trickle out into the world market, only to be gobbled up instantly by a car-hungry public at almost any price. In fact a shameful episode in American commercial life rapidly developed. Shady used car dealers bought new cars from reputable dealers and then equipped them with every imaginable accessory and sold the vehicles as used cars at prices far above normal retail. There was no legal control on what they could charge. Many veterans, some disabled, but able to drive a car, were turned away by the fantastic prices demanded.

Most of the major manufacturers were placing cars in dealer's hands and the choices available to the public steadily widened. In June 1946, Kaiser-Frazer started to build cars at the huge Willow Run plant. Although Kaiser-Frazer's plans for a front wheel drive vehicle had been abandoned due to technical problems, the new company was selling a post-war design. Studebaker's "First By Far With a Post-War Car!" (a slogan coined by Frank Tyson, a prominent Chicago advertising copywriter) was no longer a rumor, but rather a reality, as this David of the automobile industry began to take on the Goliaths with a completely redesigned line of vehicles which were to prove immensely successful. Studebaker's mechanical concepts were not revolutionary, but their styling ideas were, and these proved to be influential.

To add to the pressure on Preston Tucker, it was apparent that he did not consider the Lawson design to be mature enough for final engineering and production design, a most frustrating (and astute) posture for a man burning with ambition. Delay was debasing his challenge with each passing day. Then Good Fortune intervened. Tucker approached Tammen and Denison and was courted by Alex Tremulis, their staff expert on automotive design, to secure the Tucker styling project.

In late 1946, Tremulis met Preston Tucker for a definitive meeting. Alex walked into this meeting with a styling consultant contract in his hands, plus a generous sampling of his sketches and renderings. The meeting stretched on for hours and resulted in a marathon effort by Tremulis over many days and nights.

Tucker next came into the Tammen and Denison offices on New Year's Eve of 1947 to view the results of Alex's endeavors to capture the automotive account. As Tremulis later recalled, Preston Tucker looked over the drawings and said, "That's It!" Tucker told Alex to forget the consulting contract and to join the Tucker Corporation as its chief stylist. Tremulis agreed and thus began another phase of the Tucker saga.

Tremulis' initial design concepts did not totally refute George Lawson's work. A detailed rendering of the Tucker done on Christmas Day, 1946, was a fifty-fifty blending of the Lawson design's sharp nose, cycle fenders, garish trim and huge rear window. These were blended with concessions to reality: roll-down side windows, four full doors for passenger access, adequate space for a trunk in the front, and an engine in the rear. The cycle Fenders were retained only as a concession to Tucker, and only for the time being. (photo no. 7)

When Tremulis took over as Chief stylist in January 1947, he had no idea that many of his inspirations in that Christmas rendering were to metamorphose into reality in a matter of months. According to Alex, "At least ninety per cent of the design ideas I showed him on New Year's Eve found their way into the production car."

By March 2, 1947, the date of the Tucker '48 advertisement, the cycle fenders had disappeared, the car's contours had changed to encompass the essential space for the occupants, and the overall body had assumed realistic proportions. Yet, after all the design changes, the essence of the original Lawson model could still be discerned. In Lawson's design, the center headlight pointed straight ahead while the fender headlights turned with the wheels. In Tremulis' design, the cyclops eye became a single beacon that turned in the direction of the wheels, while the headlights in the fenders stared straight ahead. The car pictured in the March advertisement represented typical design evolution. (photo no. 8).

An essential aspect of Alex's body design work was based on his careful attention to the specifics of wheelbase, wheel tread (the lateral dimension between the wheels) and especially, internal dimensions for passengers, such as head and foot room. Also, at this point in the project, Tucker had an engine concept to present — a 589 cu. in. (9.65 l) six cylinder, horizontally opposed, liquid-cooled powerplant. This engine, still in the design stage when Tremulis joined the firm, was Tucker's pride and joy. It was to become a pivotal factor in the fortunes of the Tucker Corporation.

CHAPTER 5

THE LIPPINCOTT ENTRANCE

The March 2nd advertisement was only the beginning of Tucker's quest for a more perfect automobile. The Tremulis rendering did not represent a final design, either. This was soon to be made abundantly clear at the design offices of J. Gordon Lippincott & Company. Rumors began to circulate of an impending automobile styling account. The rumor was like a small shot of adrenaline, just enough to lend excitement to relatively mundane projects such as theater lobby popcorn machines, fountain pens, drug store counter soft drink dispensers and other useful but rather pedestrian products.

Not that we didn't occasionally bring in a light touch of our own on that 55th floor overlooking mid-Manhattan. On a particularly windy day, when the howling through the window edges reached a certain pitch, we would suspend a string weighted with a glob of clay from a ceiling fixture to just above the floor, and see how much the skyscraper tower was swaying (several inches in a real gale). On light-winded days, a lull in our intense devotion to the clients' needs often led to a bit of parachuting; not by us, but by hand-sized parachutes made of a square of tracing paper, some string and the ubiquitous glob of clay. Out the window they would go. Fascinating to track, we followed their erratic courses until they sometimes disappeared from sight, still underway. On one occasion, one of our little chutes drifted level all the way past our studio to the corner windows of Gordon Lippincott's office. He had a client in for a consultation. We never did find out if either Gordon or the client noticed that small fruit of our design genius.

Soon the rumors of the automobile styling project had a name — *Tucker*. Recalling the advertisement I had saved, I brought it out and immediately began to contrive what I thought could be done to improve the design. I had no assurance that we had actually secured the account, and if so, that I would be directly involved. I was jumping the starter's gun, and I might not even be in the race.

Although I hadn't known, the Tucker account had been under cultivation for some time. Long before the spring of 1947, Gordon Lippincott recalls,

> "Tucker gave me a ring on the phone and said that he wanted to come up and talk with me about styling an automobile. So one fine day, he stopped in the office. Tucker said that he wanted to bring out an automobile as fast as possible. I suspect that Tucker talked to other designers, but he chose us.
>
> "In our meeting (Margulies was included), Tucker agreed to cash in advance. After all, Tucker didn't even have a factory at the time. We agreed on a budget, a very modest one. I believe it was in the $40-50,000 range, [which] would be about half a million today."

Lippincott promised Tucker that his staff could produce full-size clay models directly from renderings, a factor that he now believes helped to clinch the deal. There is no record of how Tucker first learned of Lippincott, but it is safe to assume that it was through Andrew Higgins, their only known mutual acquaintance.

In 1947, when Tucker *did* have a plant and had begun the process of actually designing a car, he decided that it was time to follow through on the meeting with Lippincott and Margulies in New York and to finalize the Chicago deal. Gordon agreed to the meeting and contacted Read Viemeister in Yellow Springs, Ohio, where Read had his own design practice. When Gordon asked him if he would be interested in being a consultant on the Tucker automobile assignment, Viemeister responded affirmativly. "How could I say no? Automobile design, after my wife, is my first love!" Some years later, Read was to name his son, "Tucker."

Viemeister quickly did some sketches and renderings in Ohio and went by train to join Gordon and Tucker for the meeting in Tucker's suite at Chicago's Drake Hotel. One of Read's renderings consummated the deal. It was a simple sketch of an open car door revealing an opulent interior done as a negative rendering (white lines on black paper). He captioned it "REAL LUXURY IN TRANSIT." (photo no. 9).

That did it for Tucker. The word was, "Go" for Lippincott!

Within a few days, "the word" reached the ears of everyone at Lippincott. We had the account. Now the question was who would be on the team going to Chicago. The decision was up to Gordon Lippincott and Walter Margulies. Read Viemeister, of course, would be the principal styling consultant. Hal Bergstrom, who had been working on the Northwest Airlines interiors project, would be project manager. Tucker P. Madawick, who had been working on the Republic Rainbow interiors, Budd Steinhilber, our top stylist; and I completed the team makeup. I was in!

We found ourselves in the position of Will Rogers, the humorist and grass roots philosopher, who often said, "I only know what I read in the newspapers." We knew a little, very little about the Tucker Torpedo. It is probable that Gordon and Walter, themselves, did not know enough about the car to brief us in depth.

Before leaving for Chicago, Viemeister came to New York to do more preliminary renderings, a warm-up for the approaching event. One was a pastel sketch of a generalized Tucker '48 with an unusual front end. The bumper had a drop in the center section reminiscent of a steer horn. He took the sketch with him to Chicago.

All too soon, our preliminary work was over and arrangements made for what would probably be a two-month absence from home. We picked up our plane tickets and met at LaGuardia Airport on the morning of March 10, 1947, — destination: The Tucker Corporation.

Certainty is not generally a part of life, but the lack of it when we boarded the United Airlines DC4 flying to Chicago was a bit abnormal. We knew very little of what we would see, and had no real idea what stage the Tucker automobile development had reached. We discussed this en route, and exchanged quips about what "the car" might look like at the moment. I did a sketch on a scrap of paper of four wheels propped up by 2x4s and offered this as a possibility. No informed rebuttal was offered by Hal Bergstrom or Tucker Madawick.

Curiously, Preston Tucker had chosen to impress complete anonymity upon the designers of the two cars thus far shown to the public. At that time I did not know that George Lawson did the original design work, nor that Alex Tremulis had followed up on Lawson's efforts. I did learn something about Tremulis from Hal while on board the plane. We knew none of the details of what had been going on at the Cicero Avenue plant; a peculiar state of affairs for a car christened '48 when 1947 was already well underway. We all knew that it ordinarily takes at least two years to create a car from scratch, yet, here we were on our way to design — or redesign — the Tucker car in a matter of weeks!

After a comfortable flight (how spacious airliners were before the enlightenment of the wide-body jet age), we arrived at Midway Airport in southwest Chicago. A taxi took us to our billet and home base for two months — the splendid Southmoor Hotel on Chicago's South Side lakeshore.

The next morning, after a pleasant breakfast in the hotel coffee shop, we took a cab to our project headquarters — the Tucker Corporation. When our taxi turned onto Cicero Avenue, none of us had to bother looking for the street address, "7401." Driving past homes, stores, and machine shops, we suddenly came to a vast parking lot in front of buildings which seemed to stretch to the horizon. The taxi must have driven over a thousand feet alongside the endless facade before turning into the administration building's entry-way. Clearly Preston Tucker's dream had materialized in monumental dimensions. (photo no. 10)

Entering the reception area, we beheld a rear engine race car glistening on the shiny tile floor. It was a racing vehicle of the late 30's with a streamlined body set well within the front and rear wheels; new to us, but obviously not new. A low-slung radiator preceded a cowled cockpit, followed by a prominent engine section aft, including rude exhaust stacks which must have sounded like thunder when in action.

A placard revealed the vehicle's parentage: Preston Tucker and Harry Miller. The race car featured a Leo Goosen 400 hp six cylinder engine built in Summer 1938. It had been installed in a Miller racing car chassis for Joe Lenki, who had become a Tucker Corporation con-sultant on fuel injection and suspension. The placard stated that the car "...holds more than 30 national and international records, set on the Utah salt flats. It has been clocked at a top speed of 164 mph and averaged more than 150 mph for 500 miles." The car was entered in the 1948 Indianapolis race, but did not place.

Hal Bergstrom introduced us to the guard at the reception desk and asked for Mr. Alex Tremulis. As we continued to examine the Tucker race car, the Tucker Corporation's Chief Stylist walked the one-half mile through the corridors of the vast plant from his temporary styling office to meet us. I often wondered what thoughts went through his mind as he trekked through the halls to meet the strangers from New York.

Meeting Alex proved to be an enticing encounter. His demeanor reminded me of a debonair professor — amiable but slightly restrained, cooperative and in control of himself, but not quite confident that he was in control of what was happening all around him. Superficial con-versation and introductions aside, he escorted us back to the area of the huge plant that was to become our styling headquarters for the next two months.

During World War II this vast factory teemed with workers who turned out aircraft engines for B-29's. Now it was a hollow shell that echoed footsteps with suspenseful dramatic flourish. Many months would pass before a full staff would be recruited to run the Tucker enterprise. We walked past these empty hollows and eventually entered the fringes of the manufacturing portion of the plant. Most of the machinery had been cleared away and placed in storage, leaving great expanses of eerie, silent emptiness. It would take a highly successful venture to fully utilize such a gigantic structure and bring it back to life. The Tucker plant was awesome and intimidating — I felt small and insignificant by comparison. It was hard to fathom how even war-time aircraft engine requirements could have filled the large concrete and glass caverns. Would Tucker be able to utilize this property? It was certainly big enough to produce his touted 1,000 cars a day. With enough people he could produce 10,000 a day!

As we came into a huge assembly bay, about the size of a large air-craft hanger, the sound of our echoing footsteps was suddenly drowned out by cacophony. Once through the bay doors, the world changed. Here beat the heart of the campaign to produce the Tucker '48. Here we saw embryonic shapes in raw sheet metal coalescing into the frame and part of the body of an automobile. A nearby drop hammer pounded sheet metal from flat to contoured with ear-splitting vibrations, and the junctions of formed sheet and frame were fused under the bright sparks of welding torches. Elsewhere, men at work stations devoted themselves to the mechanical details of torching, cutting, bending, and drilling the parts of a prototype automobile. Alex gave us a cursory introduction to all of this and then led us to the design area.

A two-man crew worked on the beginnings of a full-size clay model of a car. We could discern a shape in that brownish clay which was clearly the essence of the Tucker '48 I had seen in the newspaper advertisement. I noticed that the details at front and rear were vague, without resolution.

Immediately next to the Tremulis clay model was a vacant area, a portent of things to come. Nearby was an enclosed office of about 25' x 30' (7.6 x 9.1 m), apparently Alex's design studio, now to be ours for a while. The assembly bay din was significantly diminished upon closing the door to this sanctuary. The room was small, but well lit with several drawing boards and reference tables. Here we spread out the drawings and materials we had brought with us and those which the Tucker Corporation had on hand. We awkwardly settled in and listened to what the Chief Stylist of the Tucker Corporation had to say. Tremulis was quite hospitable and gave every indication of being fully cooperative.

Our initial impression of Alex assumed new dimensions. He possessed an affable sense of humor, and chuckled frequently as he related the improbable state of affairs up to that point. He was always ready with an interesting bon mot to add to any situation, a characteristic which often aided him in his dealings with Preston Tucker.

Since he had accepted the position of chief stylist in January, Alex had brought the design of the Tucker automobile from the nebulous to the three-dimensional. He had developed a firm layout of the car which he showed us in a ⅛ size drawing, with every outside and inside dimension carefully indicated (shown on opposite page).

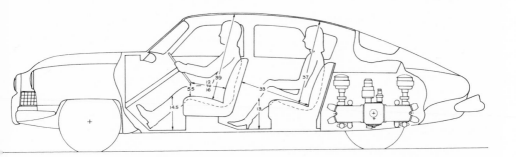

Tucker initially dictated a rather unorthodox design process. When Alex told him that a full-sized clay model of the car would take sixty days to complete, Tucker responded, "Who said clay? I've got the best metal man in the world waiting for you. Go! Go! Go!" And so, with his acute sense of shape and dimension, Alex translated his layout work directly into sheet metal. "The best metal man in the world" was Herman Ringling, a master of sheet metal fabrication in the mold of Cellini. Ringling had fashioned body parts for the 1936 Cord 810 on a drop hammer in just the same way he was forming them for the Tucker '48: "By eye!" he said. Using only the basic tools, he could pound sheet metal held in his bare hands ("Gloves interfere with your touch.") to achieve just the right shape.

Between the first of January and our spring of 1947, Tremulis had managed a crash program calling for exhausting hours of work by a dedicated crew. They brought an automobile design project from scratch to a recognizable body shape in metal without a clay model (there had been an embryonic ⅛ size clay model of little help). For a production car undertaking this bordered on the ridiculous, but he had done it. He had transferred his renderings by eye and estimated dimensions to sheet metal directly. Only in recent weeks had Tucker, for reasons known only to himself, interdicted this modus operandi and directed Tremulis to begin the full-size clay study we had seen in progress.

Up until the time of our arrival, all efforts had been directed at forging the metal marvel. Tales have perpetuated over the years that this prototype Tucker car, No. 0, sometimes called "The Tin Goose," was just a 1942 Oldsmobile reshaped in frame and sheet metal to conform to Tucker's and Tremulis' vagaries. This is hyperbole. Tremulis did analyze an Oldsmobile frame, which he considered to be the most rigid available in existing cars, as a benchmark. However, he extensively departed from it and established the required characteristics of an entirely different car. There is no record of Oldsmobile sheet metal being employed and reshaped, but a design head charged by his boss to get a car out in record time could not be very far amiss in using whatever beginning shapes were available. An existing hood or roof from another full-sized car is already contoured and worked by die forming, providing a significant advance step in achieving the designer's final results. In any event, the prototype was strictly a new creation, and our Lippincott design team would be an integral part of it.

The fourth member of our team, Read Viemeister, arrived at the Tucker plant later that first day. That evening, we met with Preston Thomas Tucker at his suite in the Palmer House on Chicago's North Shore. Tucker was accompanied by Lee Treese, a mass production expert from Ford with a long and varied background in the automobile industry, now Tucker's Vice President in charge of manufacturing. Tucker immediately impressed me as the archetypal salesman who could not only sell refrigerators to Eskimos, but also have them liking the refrigerators after the purchase. He was not a polished man, yet he had the power to interest others in his ideas. Before the night was over, I too shared the eagerness and enthusiasm of those who had chosen to follow him without reservation.

Tucker proved to be a mesmerizing champion of all the concepts of his Tucker '48. A highlight of his presentation was his description of the relationship of the car to its engine. The engine (he did not reveal that it did not yet actually exist) would "...idle at 100 rpm, drive the car at 50 mph (80 kph) at 500 rpm and would propel the vehicle to 130 mph (208 kph) at about 1200 rpm." This astounding description was not regarded by Tucker as a possibility; it was fact. He was thoroughly convinced of it. "When you step on the exhilarator, you really GO!" he said, using one of his famous malaprops.

Not one of us was an automotive engineer, certainly not an engineer of automobile engines, yet I am sure that such rpm/velocity relationships were startling to one and all. I knew a bit about boats and planes, and realized that Tucker was talking about engine revolutions per minute that were more akin to sea-going vessels and aircraft than to automobiles. It seemed to me that Tucker's background of racing cars, (Harry Miller, et. al.) was in conflict with his proposals. Race cars, especially Indianapolis race cars, scream their lungs out at 6,000 rpm and more to achieve their rated horsepower. Why was this Indy track veteran talking about 1200 rpm? He told us that Harry Miller, virtually on his death bed, had counseled, "Pres, make it (the engine) big!" Preston Tucker's engine design was indeed big. With a bore and stroke of 5" (127 mm), it was more in league with 450-600 horsepower aircraft engines. There does not appear to be any information to explain this conviction of Tucker's beyond Miller's ambiguous statement.

As the meeting steadily warmed, members of the Lippincott camp, especially Read (who had met Tucker previously), began to contribute their own ideas for the design of the Tucker '48. The subject of exhaust pipes came up, with obvious relish for Tucker, who wanted the pipes to be a sassy reminder to any car behind his that there was a powerhouse ahead. Read did a quick three-quarter rear view sketch of a '48 with three vertical exhaust pipes emerging from the top of each rear fender. Tucker thought it looked great and seemed to consider it evidence that he was in good hands. A little sober reflection squelched the concept when we all realized that those pipes would make a splendid target for mischievous pedestrians walking past a Tucker parked at the curb: Plop! in goes a rock, a banana peel, or pre-chewed gum to foul up the works.

The meeting lasted for several hours. Tucker held the floor most of the time, but Treese did his part to lend a rosy glow to the proceedings with a description of the plant's superb production facilities (it even had a magnesium foundry). Preston Tucker gave us this charge: to style the car based upon the essentials of his mechanical concepts and upon Alex Tremulis' body layout. At no point in the meeting was there any mention of our changing the fundamentals of wheelbase, wheel tread, interior dimensions or even the basic body shape. It was a classic case of pure styling. The primary dimensions were inviolate, as were the tapering roof, and of course, the cyclops eye front and center (we were to have fun with that feature). Outside of these requirements, there were no constraints. We were expected to go all out in our efforts.

As our Lippincott crew taxied back to the Southmoor that night, we carefully considered the magnitude of the task that lay ahead. We already knew of the primitive nature of our basic set-up back at the design area. We had to start, almost from scratch, on a valid and professional presentation which would involve a second clay model right alongside the one recently started by Alex Tremulis. We realized that, with each stroke of our clay tools, we would be competing with whatever evolved next door. We were the new kids on the block and everybody would be watching us.

CHAPTER 6

DOWN TO WORK

Back at the plant the next morning, project leader Hal Bergstrom assembled us together in our design studio to establish a project plan. Our "design studio" was just an enclosed space in the factory with a few drafting tables for our sketches and renderings; plain, but adequate. Papers and pencils were at hand, and we had a supply of black railroad board and Prismacolor pencils for those striking renderings in negative which George Lawson had pioneered years before. This was the dramatic and quick technique Read Viemeister used in the white-on-black "Luxury in Transit" rendering which so impressed Preston Tucker some months before.

We had ideas boiling over, just waiting to get onto paper and board, but we needed to ascertain our next steps. Gordon Lippincott had told Tucker that the time constraints did not allow for ¼ scale clay models, and that the procedure would have to be direct translation of design concepts to full size. A two-pronged effort was called for: produce a raft of valid two-dimensional ideas while the foundation for another clay model was being completed.

The foundation for a clay model is the "armature" or "buck." A sculptor designing a statue, bust or monument will frequently do a study in oil-based clay first. This may be carried through to finished detail and a casting made in plaster in whole or parts, or measurements made to transfer the clay model to another medium. Unless the study is very small, it cannot be solid, since clay is expensive and not structurally strong. Neck, arms and legs for a statue, for example, must be supported inside with a strong material (wire, rod or wood) — the armature. The clay often used is known as Plasticene which is perfect for the master of a plaster cast. In automotive modeling, a somewhat different clay is used, which is first heated in a container and then applied to the armature. This sulphur-based brown clay becomes very hard when cool, and can be given a finished surface smooth enough to paint for a final realistic presentation to management.

The armature or "buck" for an automotive model is generally of wood, a hollow structure smaller than the expected automobile will be. It is usually constructed of wood lath nailed to plywood and lumber supports propped up as inconspicuously as possible. (photo no. 11) Clay is then applied, troweled on to reach the approximate anticipated shape. It is then sculpted by clay carving tools, templates, straight edges and no small amount of direct human intervention.

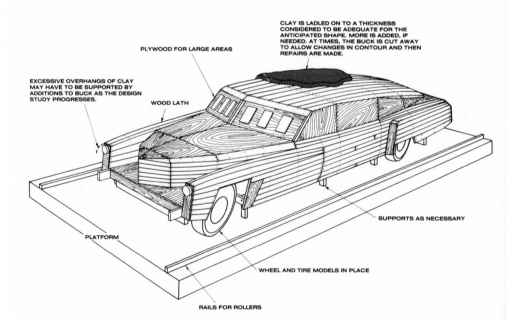

CLAY IS LADLED ON TO A THICKNESS CONSIDERED TO BE ADEQUATE FOR THE ANTICIPATED SHAPE. MORE IS ADDED, IF NEEDED. AT TIMES, THE BUCK IS CUT AWAY TO ALLOW CHANGES IN CONTOUR AND THEN REPAIRS ARE MADE.

PLYWOOD FOR LARGE AREAS

EXCESSIVE OVERHANGS OF CLAY MAY HAVE TO BE SUPPORTED BY ADDITIONS TO BUCK AS THE DESIGN STUDY PROGRESSES.

WOOD LATH

SUPPORTS AS NECESSARY

PLATFORM

WHEEL AND TIRE MODELS IN PLACE

RAILS FOR ROLLERS

The finished clay model serves the primary purpose of being "That's it!" or "needs work," (or "yech!") to the assembled throng of management, marketing, sales and other executive and administrative persons who pass judgment on automobile designs. If accepted, its usefulness is doubled; it becomes the master for the exterior contours of the projected vehicle.

To accomplish this, the original buck is installed on a level platform with parallel tracks. A bridge is installed which surrounds the model and can be moved on wheels fitted onto the tracks. Dimensions are applied alongside the tracks which match a fiducial (pointer) on the bridge so that the fore and aft positions can be logged. The bridge is fitted with holes up its sides and across the top that aim in toward and down on the model. These holes are located at dimensional increments that can also be logged. To determine a dimension, a marked rod is fitted into the holes and carefully moved toward the model just until initial contact. Thus, a reading of the cross-sectional contour of a car model at station 10 (ten inches aft of a set datum, or start) is made by setting the bridge at 10 and inserting the rod at each hole, recording the distance the rod traveled in the bridge to the point of touching the

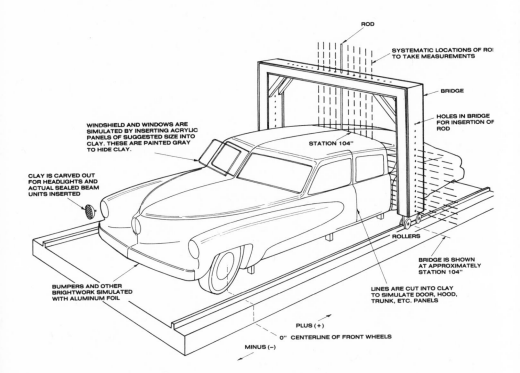

ROD

SYSTEMATIC LOCATIONS OF RO
TO TAKE MEASUREMENTS

BRIDGE

HOLES IN BRIDGE
FOR INSERTION OF
ROD

WINDSHIELD AND WINDOWS ARE
SIMULATED BY INSERTING ACRYLIC
PANELS OF SUGGESTED SIZE INTO
CLAY. THESE ARE PAINTED GRAY
TO HIDE CLAY.

STATION 104"

CLAY IS CARVED OUT
FOR HEADLIGHTS AND
ACTUAL SEALED BEAM
UNITS INSERTED

ROLLERS

BRIDGE IS SHOWN
AT APPROXIMATELY
STATION 104"

BUMPERS AND OTHER
BRIGHTWORK SIMULATED
WITH ALUMINUM FOIL

LINES ARE CUT INTO CLAY
TO SIMULATE DOOR, HOOD,
TRUNK, ETC. PANELS

PLUS (+)

0" CENTERLINE OF FRONT WHEELS

MINUS (−)

model. Longitudinal readings are made the same way, keeping the rod in the same hole, say at centerline, and incrementally moving the bridge the full length of the car — or just a fender or a hood — and repositioning the rod with each relocation of the bridge.

It might be interesting to note that in 1942, my primary assignment at Edo Aircraft was the use of a new measuring process which was a peek into our computerized future. Using an electric calculator and a mathematical formula, I lofted the shape (drew the lines) of a 500″ (127m) long aircraft float (pontoon). At each station or bulkhead from the bow to the stern, I calculated the profile required to achieve smooth flowing lines. Edo was one of the aircraft companies pioneering this method in the late 30's. With a computer, instead of a calculator, I could have registered each result of my calculations onto tape or disc and been able to "draw" the shape automatically — as is done today.

Automobile design now enjoys the benefits of modeling new designs on a computer screen. The graphics computer now allows visual concepts and dimensional readings to be made *instantly*. However, the automated and the mechanical approach are the same in the axiom: "garbage in, garbage out." Talent and creativity design cars, not computers, and the ultimate aim, by any method, is a full-sized model to see, and a working prototype in which to sit and drive. The mechanical bridge used in automobile design was invented by Gordon Buehrig, the designer of the Auburn Speedster, the Cord 810/812 and numerous Duesenbergs.

The Tucker Corporation had a pattern shop, recently set up, and a number of able hands did the buck's carpentry. Alex Tremulis and Tucker Madawick were both experts in the techniques of clay model bucks, platforms and bridges. Read Viemeister and Budd Steinhilber had excelled in their studies at Pratt Institute, which included much clay modeling. Read had also acquired first-hand experience at Graham-Paige on the earlier Lippincott project there. Hal Bergstrom was a seasoned veteran of most design aspects.

My own qualifications were more limited, but by age twenty-six, I had had some experience in clay work. I felt I had a vivid imagination, a strong familiarity with model making, and an eagerness to learn and to get my hands on a genuine automobile project. This was it.

Everyone pitched in, both on the drawings and the planning for the clay modeling ahead. By now Budd Steinhilber had joined us and the Lippincott crew was five in number. Every detail had to be quickly worked out in order to compress the elapsed time. We needed some advanced decisions such as confirmation of wheel and tire sizes. The pattern shop made them out of laminated wood painted black with white sidewalls. These had to be in place on the buck before any clay was applied.

It was not long before Preston Tucker began regular visits to discuss our sketches and renderings, and to monitor the progress of the buck. At this point, we began to fully comprehend what unusual conditions prevailed in the design area. Inside our studio, we enjoyed a definite degree of security. For now, we were free from the gaze of curious passersby, but this situation would endure only until the day we started on the no. 2 clay model. Then, unlike any design team we had ever seen or heard of, we would be out in the open for anyone in the development hanger to come over and have a look at what we were doing. And if we knew anything about human nature, their curiosity would not be accompanied by silence, either at the time of their visit or later, when the visitor spread the word to their compatriots about what the New York gang was doing.

Our daily commute from our quarters in the Southmoor to our compact studio in the vast reaches of the Tucker Corporation plant was sometimes by taxi or often by a company car chauffeured by the son of Ralph Hepburn (Hepburn Sr. was one of the most distinguished race car drivers in history). Through interesting conversations with this young man whose acquaintance with cars probably began the day he first learned to recognize words, we began our immersion into the world of the automobile.

Ralph Hepburn was a technical consultant to the Corporation, and as the regional manager for the West Coast, he was being groomed for leadership in prospective company racing programs. These programs included captainship of a team which was to pilot Tucker racers in the 1950 European Grand Prix. Hepburn was tragically killed in 1948 while practicing for the Indianapolis race of that year. His death ended all of the Tucker racing plans.

With the work on our wooden buck, platform and bridge nearing completion, we were able to concentrate on the paper-sketching and board-rendering phases of the project. We ran off reams of sketches and numerous renderings to illustrate our proposals. Hal Bergstrom, our project manager, proved to be just the right balance of scout master (getting time cards and expense accounting in on schedule for dispatch to headquarters in New York) and hands-on participant. Tucker Madawick alternately sketched valid concepts and brooded over the poor quality of the clay we would be working with, how we were going to heat it properly, and how the entire set-up compared to his experiences at the Ford Motor Company (unfavorably, of course). Read Viemeister was a fountain of ideas, as was Budd. And I, absolutely enthralled by it all, was constantly congratulating myself for having the good luck to be smack dab in the midst of an automobile design project. Actually, there wasn't a phlegmatic personality in sight.

We were designing a car at a time when chromium and stainless steel trim was an increasing trend, and yet we all agreed that the car's fundamental *shape* should be the key to the Tucker '48's identity. Within the confines of our charter, shape could not include drastic changes in the mid-body of the vehicle; in the details of front and rear, however, there could be dramatic breakthroughs. According to Preston Tucker's instructions, the Tucker '48 should present a striking image as it approached and a dramatic impression as it passed. This stimulated us to experiment with boxy shapes, rounded rear ends and elliptical contours, all the while avoiding excessive trim. It wasn't easy at that time.

This was the era of the *radiator grille,* for example. The 1920s radiator —which radiated heat directly, a rather honest statement — had died in the early 30's and had been supplanted by die cast and stamped metal facades which often looked like either waffle irons or bird cages. A few 1937 Ford owners even adorned their grilles with tin birds resting within the open work. Pre-war French Peugeots had grilles lacy enough to allow the headlights to be mounted within.

The Tucker challenge was not made any easier by the news that Alex and the Tucker engineers had not yet answered a particularly crucial design question: where would the radiator reside? They did not know if it was going to be in the front or the rear of the vehicle. They had no concrete evidence, no wind tunnel tests upon which to base a decision — absolutely no empirical data. They did have plenty of ideas, some of them quite exotic. One called for a fin-and-coil radiator at the lower front of the car with cooling hoses running in the rocker panels (the area below an automobile's doors), conveying coolant to and from the engine in the rear. Another had the radiator located in the rear of the engine compartment, with air intakes in the forward edges of the rear fenders. If the latter were done, the radiator would probably be placed aft of the engine up against a grille at the extreme end of the car.

It would be some time before the final returns were in on this issue, but we had to be prepared for either eventuality. In the meanwhile Preston Tucker insisted that the flowing lines of the vehicle could not be disturbed; a flat front end to capture air for a forward radiator was definitely taboo, as were air scoops to capture and direct air toward a radiator in the rear. To cool the rear engine, there had to be air intakes in the rear fenders and a large grill across the back of the Tucker '48. Further, we were told that the front end could not be bulbous above the bumper line, and the sharp cleavage of the Lawson and Tremulis designs must be retained. We were designing a dichotomous vehicle.

A rather interesting aspect of the project developed due to the lingering presence of a very imposing gentleman — William Stampfli. He had the bearing of an Italian count, the physique of a wrestler, the smile of a guru and the knowledge of a master engineer (his title was "chief mechanic," for some strange reason). He visited our design site quite frequently, always making succinct comments about our progress, with erudite references to the stages of development going on elsewhere. One day, I walked with him over to the metal prototype site (it was still being worked on, but at a slackened pace awaiting design direction) and he asked me, "Well, Phil, How do you like it here?"

"It's great. Marvelous to be a part of such an exciting assignment."

At that time I was extremely curious about the engine to be placed into "our" car, and I asked, "Mr. Stampfli, what's the status of the engine program?"

"Oh, we hope to have the castings for the crankcase and the cylinders this week," he replied. "It's coming along quite well. Terrific engine. Tucker's latched on to one of the best engineers in the business. He's come up with something that should be just right."

'Castings this week', I repeated to myself. "What engine?" he should have said. The Tucker Corporation was raising funds for an automobile which didn't yet have an engine. Here they were, building a sheet metal prototype which would ultimately be patterned after the clay models in process, and they still had no means to make the car go. It was the second quarter of 1947; orders were being taken for delivery of production cars in 1948. The dawn came up. It wasn't just *us* who were in this situation; the engine crew was part of the drama too. The impact of this revelation was to be felt for a long time.

I never did learn the name of that "best engineer in the business." The name "Jimmy," (possibly Jimmy Sukayama, Department C-41, "Special Assignments") was often heard, as was that of Ben Parsons (director of engineering), as the father of the engine, but we in the design bay were never quite sure.

The question of why Preston Tucker chose a huge low-speed power-plant for his dream car aside, the existence of a proven engine and drive train was essential for the success of the automobile. The lack of a suitable engine had doomed the L-29 Cord, dulled the vitality of the Lincoln Zephyr and cast a pall of mediocrity on the Kaiser-Frazer cars (which were originally envisaged as high performance front wheel drive vehicles).

The revelation by Stampfli assumed greater consequences when I listened to Alex's briefing on the current body design. The car's configuration included provisions for the Tucker engine in the rear, but without the vaunted hydraulic pump and hydraulic motors in the wheels so crucial to *"Flowing Power."* At some point between the first of January and our entrance into the project, Preston Tucker learned the facts of physics which ruled out such an idea. I don't know how he learned the facts, but most of his engineers knew that it was not possible, in 1947, to propel enough hydraulic fluid at sufficient pressure to remote motors to drive an automobile with adequate acceleration and at road speed. The power to do so would have been prodigious.

As it was, Tucker's engine design was unusual for its time. He planned to place the huge engine transversely with large fluid drive housings at each end of the engine crankshaft. These were to be connected to drive shafts going to each of the independently sprung rear wheels. Six individually cast cylinders were grouped in two trios on each side of the crankcase, a coolant manifold fed each cylinder in turn. Individually cast cylinders had not been common for liquid-cooled engines for many years. The engine had hydraulic lines to each of twelve intake and exhaust valves, plus a large generator and a mammoth starter. This was, indeed, a novel engineering concept for the post-war era.

These engineering revelations would have been ominous in the absence of Preston Tucker's persuasive personality. His ideas and his charisma entranced countless people, many with imposing credentials. Could he somehow carry the whole thing off? It was certainly worth seeing through.

1. EXCALIBUR: The four-door sedan version of the author's proposed 1944 "car of the future."

2. Read Viemeister working on the clay model of a proposed Graham-Paige post-war car.

TORPEDO ON WHEELS

LONG, LOW, and streamlined, the Tucker Torpedo has doors that extend into the top for easier getting in and out. Brakes are of an automatically adjusting type developed originally for racing cars. A new engine can be installed in 15 minutes when necessary

Engine in rear, all-hydraulic drive

make the Tucker a real car of the future

MORE like a Buck Rogers Special than the automobiles we know today, the Tucker Torpedo is scheduled to hit the road sometime in '47. If all goes according to plan, this startling car will incorporate a series of spectacular engineering innovations that conservative auto manufacturers have classified as "at least five or six years off."

Here are some of the highlights of this 126-inch wheelbase vehicle in its present design:

Hydraulic torque converters provide a direct power-transmitting system that does away with the customary clutch, transmission, drive shaft, differential, and rear axle. That eliminates about 800 working parts and saves

OPERATION of car's hydraulic drive is shown in diagram above. Fluid in reservoir (A) flows to pump (B) driven by car's engine. Pump forces fluid through flexible pipes to hydraulic motors (C), which drive rear wheels. Fluid then returns to reservoir for re-use.

SCIENCE ILLUSTRATED

3. The December 1946 "Tucker Torpedo" advertisement in **Science Illustrated,** gave the public its first look at Preston Tucker's dream car.

Now it can be told...

HOW 15 YEARS OF TESTING PRODUCED THE SURPRISE CAR OF THE YEAR

Here's the Success Story of America's Newest most exciting Motor Car

YOU'LL BE SMART if you wait! For this is IT ... a car completely new, yet with engineering principles completely proved...the SURPRISE car of 1947.

Already pilot models that will be the inspiration of engineers for years to come are now being completed in the TUCKER plant, largest and most modern in the world.

Already preparations for the most advanced tooling and production lines are being completed.

Already Tucker distributor and dealer franchises in some major cities and states have been awarded.

Later in 1947 you'll judge the new car for yourself. Be among the first to give the Tucker '48 a workout.

WHEN YOU DO, THIS IS SURE:

You'll get the motoring thrill of your life—you'll find nothing you've experienced before will compare with the smooth surge of FLOWING POWER from the Tucker *rear engine drive* ... the new comfort of riding on the unique Tucker individual wheel suspension ... the feeling of security you have in driving a car so precisely balanced that it almost drives itself.

You'll discover the completely new car you've been waiting for so many years is not just a postwar dream ...but really here...five years ahead of its time, yet with engineering principles *completely proved* by fifteen years of rigid tests.

For Years, the famous engineering team of Preston Tucker and the late Harry Miller designed special cars, with engineering features years ahead of those turned out in volume production factories. In fifteen years, Miller Specials won eleven of the annual Speedway Classics at Indianapolis, recognized as the greatest testing ground for automotive progress.

At Indianapolis, the pit of the Miller Special was always the center of interest for manufacturers of both racing and passenger cars. But many of the features developed by Preston Tucker could not then be utilised in mass production cars. So they were stored away, for the day when volume production methods would be advanced to the point of utilising them.

When War Came, Preston Tucker went to Washington and, from his storehouse of tested designs, developed engineering features in the motorized vehicles and aircraft that helped win the war. His ideas went into bombers and pursuit planes ... into the Sherman tanks that defeated Rommel...and into the motorized artillery which spearheaded Patton across France.

Challenged by War, American industry developed new production methods. At last, the new engineering ideas of Tucker would be volume-produced. So, when peace came, the great B-29 plant—largest and most modern in the world—was found to be the best to turn out the Tucker '48 ...the first completely new design in motor cars for fifty years, with engineering principles completely proved.

Now, at Last, Tucker is on its way. With finest tools and dies, plus an organization of top motor executives—Hanson Ames Brown, former General Motors vice president; Fred Rockelman, former president of Plymouth Division of Chrysler; Lee S. Treese of Ford; Chief Engineer K. E. Lyman, formerly of Borg Warner and Bendix, and other leading motor car executives.

WHEEL-BASE 128 INCHES

THE NEW TUCKER...YEARS AHEAD!

Now FLOWING Power. Flowing power —sure as a mighty stream—moves direct from engine to wheels through hydraulic torque converters. No conventional transmission or clutch.... or conventional differential, either. So the power is smoothed out for a ride free as a sea gull's glide.

Single Disc Aluminum Brakes. No conventional brake bands to wear ... no adjustments to make, *ever*. These utterly new type hydraulic, air-cooled brakes are 63% more effective ... the same type that stop fast-moving planes on a carrier's deck without skid or turn.

Precision Balance. The unique Tucker design distributes weight to give maximum safety, maximum power transmission and hairline steering and driving control. Only a rear engine drive can achieve this precision balance—for years the goal of all automotive engineers.

New Individual Wheel Suspension. The NEW Tucker suspension makes conventional spring suspension obsolete. Each wheel is cushioned by its own resilient action arm, actually eliminating shock instead of simply softening it. Tested for years on America's Speedways.

Rear Engine Driv. Bird's eye cutaway view of the Tuckerflat opposed 6 cylinder engine. Electroni high frequency ignition, measured oil injection. Higher power-weight ran than any volume production automotive engine ever built. This means greater eonomy, flexibility, faster starting, troublee operation, safety.

Tucker '48

COMPLETELY NEW Yet with Engineering Principles **COMPLETELY PROVED**

Address inquiries to Executive Department TUCKER CORPORATION 7401 South Cicero Avenue, Chicago 29, Ill.

Philip S. Egan

4. The March 2, 1947, "Tucker '48 advertisement featured in many national newspapers including the "New York Tribune," showed the evolution of the Tucker automobile.

5. The 1946 Tucker Torpedo was a one-quarter size model skillfully crafted by George Lawson and presented as the real thing.

U.S. AIR FORCE

6. This Boeing Dyna-Soar — forerunner of the Space Shuttle — evolved directly from designs by Alex Tremulis developed at Wright Field in the mid-1940's.

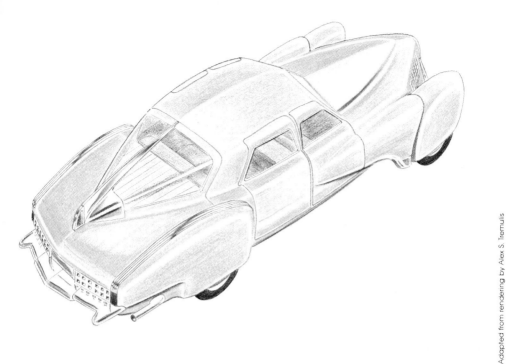

7. Alex Tremulis' initial design conception for the Tucker Torpedo was a blending of George Lawson's sketches and Preston Tucker's ideas, tempered with automobile design reality.

8. Alex Tremulis' spring 1947 rendering of a Tucker '48 set the theme for the no. 1 clay model.

Real luxury in transit!

9. Read Viemeister's elegantly simple rendering, "Real Luxury in Transit!" helped to cinch the Tucker account for J. Gordon Lippincott & Company.

Labels on image:
6. FORGE PLANTS AND DIE SHOP
4. ALUMINUM FOUNDRY
PULASKI RD.
3. OIL AND "CHIP" SHOP
5. MAGNESIUM FOUNDRY
8. PRIVATE DRIVEWAYS INTO PLANT
2. POWER PLANT
MACHINE SHOP AND ASSEMBLY PLANT
7. PARKING LOT
CICERO AV.
1. ADMINISTRATION BUILDING

Richard E. Jones collection

10. An aerial view of the Tucker Corporation plant. The former Dodge aircraft engine plant was one of the largest factory buildings in the world in 1947.

11. Alex Tremulis (wearing jacket) supervises the application of clay to the wooden buck soon to become the no. 1 clay model.

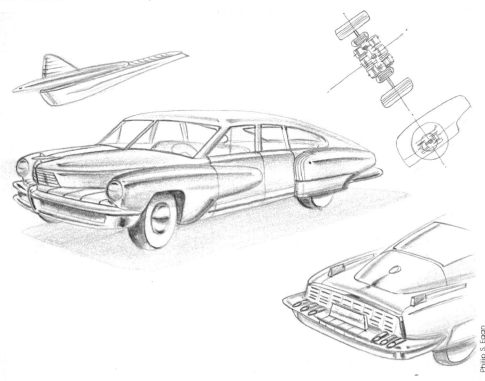

12. Pencil sketches of progress made on the Lippincott no. 2 clay model from the author's pictorial diary of the Tucker project.

13. The Lippincott clay model nearing completion. Note the parking lights on a wide horizontal band above the bumper which has been covered with aluminum foil to simulate chrome. The wooden bridge is seen surrounding the rear half of the model.

14. A rear view of model no. 2. The rear window was too high, and exhaust pipes were set into the foiled bumper.

15. By mid-April, 1947, the rear engine air intakes had been refined, and the rear quarter window had been extended back to provide more visibility.

16. The front grille contined to evolve and the vertical bars vanished. Acrylic placed in the windows and windshield added realism.

17. Without ever leaving the starting gate, the Tremulis (foreground) and Lippincott models race toward the finish line.

18. Putting the finishing touches on the Lippincott model are, at left, project manager Hal Bergstrom, Budd Steinhilber, and Philip Egan; with his hand on the door handle is Walter Margulies, Gordon Lippincott's partner.

19. During the last weeks of the project, the Lippincott team obtained Tucker's permission to convert the left side of the model into a more advanced 1950's design. Kneeling in front of the car is Budd Steinhilber, behind him is Hal Bergstrom, the author stands at the windshield, and Walter Margulies is at the rear door.

20. An elated Preston Tucker waves to the assembled crowd of over 5,000 as he and the "Tuckerettes" show off the prototype Tucker '48, on June 19, 1947.

21. The Tin Goose — long, low and powerful — its message was one of austerity.

22. Read Viemeister's steerhorn bumper gave the car a dramatic grace, even when it was still a painted wood mock-up with metal inlays.

23. The prototype effectively blended the body contours of the Tremulis clay model with the front and rear ends proposed by the Lippincott team.

24. The driver's control layout was a poor specimen, hastily cobbled together from the instrument cluster of a derelict Studebaker. It did convey Preston Tucker's passion for a minimum of controls, but this area needed much more attention.

25. The horn button design for the proposed safety steering wheel. Its contour blended with the hub to provide a relatively large and smooth surface.

26. "The Big Engine That Couldn't." The Tucker 589 engine. This engine powered the first working prototype, but proved ineffective for production models.

27. "The Smaller En-
gine That Could."
The Tucker 6ALV-335
Air Cooled engine.
Carl Doman called
it, "a fine operating
power plant."

CHAPTER 7

ON TO CLAY

Sculpting in clay is a wonderful experience. From a mass of stuff ladled onto an armature or buck, tools and hands cut and shape what the mind's eye indicates. A sculptor can be let loose on a bulk of clay and come forth with a face or a bird or a fender.

We did just that. A pile of sketches and renderings gave us a repository of ideas to convert from flat to three dimensional. Tucker looked over some of our paper and board conceptions and liked one here and one there. These we scheduled to render in clay. We privately kept other images for future consideration.

Our wooden buck was at last ready. We plastered hot clay onto it in abundance, and then, guided by Alex's layouts for basic body contours, we set our sights on a completed preliminary car form. From this, we could begin our excursions into design variation.

Tucker Madawick had devised a simple wooden box with light bulbs inside to provide heat to soften the globs of clay before applying it on the buck. His worries about the quality of the clay were well-founded. It was not really first class material. It was too soft and its color was more gray than the vivid brown he had been accustomed to at Ford. But it was what we had, and we accepted it, hoping that it would harden enough to provide a good carvable surface.

The clay tools used in automotive design are big brothers to those used in sculpting a bust or a bird. Some large clay tools can be purchased as-is, in art stores, but often it's necessary to make tools for large surfaces by adapting hacksaw blades or straight edges of plastic or wood to individual requirements. Ideally, it is useful to have "sweeps" of plastic or metal which have been cut accurately to very large radii — several hundred inches. This is advisable because the sheet metal of an automobile will tend to "oil can" (flex in and out when pressed against) if it does not have sufficient convex (outward) shape. This is not easy to avoid in sculpting large areas, such as a roof top, by eye. Good sweeps will act as a control of this problem. We had a bare minimum of such frills. We all felt a little like the cartoon character of the sculptor in his studio, standing before a block of granite, a simple chisel and hammer in hand, saying "smile" to his model standing nearby. However, it didn't take more than a few days before we had a basic shape to sculpt.

Our area alongside the Tremulis Tucker was certainly no studio. The gray walls and concrete floors of the plant lent a construction-site feeling to the surroundings. Natural light came in through high windows, and artificial light came from white reflectors surrounding floodlight bulbs in the ceiling. We immediately asked the plant carpenters to erect portable partitions as back-drops, about six feet high, on which we could pin our sketches and renderings to lend color and practical reference material to the proceedings.

It was now April; we had been on the job about three weeks, working at full bore, putting in at least six day weeks, with much overtime. We were enjoying every minute of it. Preston visited the site several times a week, dressed in his three piece suit, often wearing a homburg and sporting a cane with which he often pointed to some feature of the model. He was frequently accompanied by an entourage of Corporation associates, some of whom we began to recognize. Stampfli was one of them. He generally had more intelligent comments than most, and he often lingered after Tucker departed to make sure some of his observations sank in.

Tucker's entourages also included signed-up and would-be dealers who received a package tour. First they would see the metal prototype (if the drop hammer was in action, the racket made the scene more dramatic), and then see men doing various mechanical functions at work stations in the development bay, and then a four cylinder fuel-injection engine mounted on blocks about thirty-five feet (ten meters) from our clay model. The engine had been designed and built by some of Tucker's racing associates. He loved to fire it up (it had no muffler) and run it at high speed for a few minutes. When the exhaust smoke became too choking, he would cut its throttle and beam at the admiring throng with those big brown eyes of his.* The entourage would then converge on the clay models. Preston would point with pride at various sculpted features, and describe what we were doing.

Some of these visits were at decision-making stages in the no. 2 clay model. Tucker would arrive with at least two associates, and we would refer to the drawings on the backdrop and then to the area of the car which corresponded. Budd Steinhilber always thought of it as picking from a Chinese menu, dish after dish, idea after idea.

Gordon Lippincott came to Chicago shortly after we finished covering the entire wooden buck with clay and began shaping it. He was obviously pleased with our progress. "In a matter of weeks, you guys had gone from sheets of paper on the walls to pushing clay" he told me later. At this stage we had only a rough shape of the Tucker '48, and were well aware of the long road ahead. Gordon was very excited with the project, and became as involved as the rest of us. He had brought some periodical photos of a Convair XB-46 twin-jet light bomber which had just been test flown. The jet's air intakes were elliptical in shape, very avant-garde, and he thought that the theme might be worth trying on the front end of our clay model. We absorbed his photos and their impact, and set to work, adding globs of clay to the front end to support the concept. Roughed out, the idea did, indeed, have merit, but then a funny thing happened. Preston Tucker came by, drank in the whole concept and politely said that it did not abide by his intentions. In his view, the front end must include the Cyclops eye at top center and provide for air intake at the bumper level, period, paragraph. So much for elliptical excursions. In fact, Gordon expressed the opinion that the Cyclops eye alone was enough to make the Tucker automobile unique. Perhaps he didn't know of the long history of center-mounted headlights in Europe; or perhaps he was simply a prac-

---

*Robert F. Scott, in "Ordeal By Trial" published by *Automobile Quarterly*, Volume 2, no. 4, 1963, states that Andrew Higgins called Preston Tucker "the world's greatest salesman. When he turns those big brown eyes on you, you'd better watch out."

tical businessman who knew that, ultimately, the client had to be satisfied.

It had been a long day, that of the elliptical front end effort. We joined Gordon for dinner at a roadside restaurant on Cicero Avenue, where, on the boss' tab, we had a good meal and much frank talk over the din of the then-ubiquitous juke box. Though lost in the recent war, Glenn Miller could still be heard in Moonlight Serenade.

Our regimen of home base at the Southmoor Hotel and work at the plant, became an exciting routine. The "routine" was late dinner at a nearby restaurant on the South Side, sleep in our large suite, which the hotel had converted to communal sharing, and then an early breakfast in the hotel coffee shop. The "excitement" was this fascinating interlude in our lives as transients in a community of automotive devotees brought together through the determined genius of one man, Preston Thomas Tucker.

Another genius, Alex Tremulis, had an impressive knowledge of automobiles and their design. Though obviously troubled by his forced competition with us (his clay model side by side with ours), his stories about various automotive projects were fascinating. His sense of humor was a marvelous source of entertainment. We often had occasion to ride with him in his right-hand drive Lincoln Zephyr convertible. Budd recalls that during one ride, Alex "hunched down behind the steering wheel, and I hung out the left window, hands flailing in the air, much to the consternation of adjacent travelers who must have assumed that I was driving the car."

Alex loved to tell jokes about himself, such as the time he was crossing a bridge in an open car and lit a cigarette with a prized Zippo lighter — and then threw the lighter into the river like a spent match. His descriptions were always sparklingly vivid. He spoke of a car design "so low that you'd have to reach up onto the curb to strike a match."

Among the friends of Alex Tremulis and other Tucker personnel were the vivacious Granatelli brothers. They had an auto repair shop in Chicago where Alex went to have his temperamental Zephyr V-12 worked on. The medium-sized Granatelli operation became the headquarters for a variety of enterprises related to street and race cars. One of their bright ideas was a specially formulated additive/lubricant poured into the crankcase of an automobile engine to improve performance. They called it STP — the Granatelli brothers' slick road to riches. The STP logotype is possibly the most enduring of all sponsor logos found on race cars to this day.

As we became acquainted with the Tucker Corporation staff, an overall characteristic became apparent: they were more maverick than opportunist. Each one had broken away from traditions that he or she no longer accepted. Most of them were experienced veterans of the automobile industry; in no way were they castoffs. Rather, they appeared eager to adventure into realms different from what they had experienced before. This took courage. It would have been much easier for most of them to simply phase back into post-war life in Detroit or wherever they had been before. More than once I detected inferences that they were not highly regarded by their former colleagues, simply because they had been more adventuresome.

The roll call of these courageous souls was impressive, and included:
Fred Rockelman: Tucker Vice President and Director of Sales.
> He was formerly the President of the Plymouth Division of Chrysler, and General Sales Manager of Ford.

Hanson Brown: Executive Vice President for Tucker Corp.
> Formerly a Vice President for General Motors.

K.E. Lyman: Development Engineer.
> Top engineering consultant and former Bendix and Borg-Warner executive.

Ben Parsons: Vice President and Chief Engineer for Tucker.
> International expert on fuel injection.

Lee S. Treese: Vice President in charge of manufacturing.
> Formerly a production executive at Ford.

Herbert Morley: Director of Materials for Tucker Corp.
> Former plant manager from Borg-Warner.

Robert Pierce: Vice President and Treasurer.
> Formerly secretary-treasurer of Briggs Manufacturing.

Other men of note who were caught up in Preston Tucker's dreams included: Al McKenzie, a racing mechanic of the Horace Dodge boat team; Eddie Offutt, a former race car driver, and top mechanic who became Tucker's Chief Engineer in Engineering; Gene Haustein, one of the best test drivers; Joe Lenki, a racing circuit driver of note; Stampfli,

Ringling, Tremulis and many others. They all brought a wealth of knowledge and dedication into the Tucker Corporation.

Prior to the spring of 1948, there was also the presence of racing great, Ralph Hepburn, who often visited headquarters for business meetings with top management. He was an accessible celebrity for anyone who wanted to talk to him. On a number of occasions Hepburn chaired meetings with such racing notables as Rex Mays, Johnny Parsons and Mauri Rose. Around one of the large tables in the company dining room they would discuss strategy or policy regarding a threatened drivers' strike or new safety regulations at Indy.

Improved plant facilities, such as the company cafeteria and dining room, an assembly area under construction and the gradual evolution of amenities, were often noted on long walks past incredible expanses of unused space or going via tunnel to an adjoining building. Only a few months before, these reaches had been thriving thoroughfares of countless workers building military hardware. Now, an ever-increasing number of men and women was beginning to bring the place to life again in a more constructive pursuit.

The spring of 1947 was a great scene.

CHAPTER 8

FOCUS AND REFLECTION

Preston Tucker's negative response to our elliptical front end, was galvanizing, and excellent discipline. Every project, individual or team, needs a central theme to focus on; to do otherwise is chaos. We proposed a modest departure from Tucker's original charge and he said "no". We were still exploring the limits of our permissible excursions. Clients are usually not designers and therefore cannot be expected to understand what might yet unfold. The Cyclops eye was a logical target for our next efforts on the front end, and that just-in-case grille at the rear was a concurrent goal.

Tucker's idea of a fixed circular headlight lens behind which a General Electric sealed-beam headlamp would pivot with the steering seemed non sequitur. Why not design a panoramic lens that would transmit the rays of the headlamp evenly? We began studies of this idea, blending the rest of the front end into a graceful shape immediately above the bumper-level grille prescribed by Tucker. We also worked in earnest on the bumper/grille to make the entire front end an integrated image. At the end of two weeks of steady work on the clay model, we had a complete car, quite rough in spots, but an expression of an entire design theme. The panoramic lens appeared to fit in well with the car's overall aspects.

At this stage of the project, I felt that significant steps had been made which should be retained. I began a pictorial diary to record the evolution of the Tucker (photo no. 12). The diary was entrusted to

Author's sketch of the panoramic lens

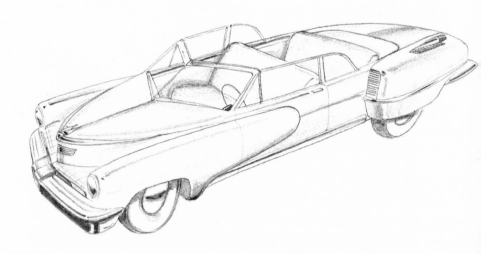

Author's idea for a Tucker convertible.

relatives, John and Dorothy Ristine of Evanston, Illinois, who kept these quick sketches for almost forty years until I remembered their existence. One of these sketches was a purely fanciful Tucker convertible.

The rear end of the car began to look good. However, the air intake on the rear fender looked like pre-war styling ("needs work", we said to one another). Still, many of Read's and Hal's ideas were distinctive and successful. The exhaust pipes had been moved to the top of the bumper, and Preston now had that sassy rear end, replete with an abundant grille to take in or let out air — in the event that the radiator was going to be in the rear. Tucker Madawick's and Budd Steinhilber's creativity and skills in clay modeling had shortened the time required to get this far. The results were those of quality.

Tucker continued his visits, occasionally alone, but usually with a retinue (the area was practically a public forum anyway). He was generally pleased with what he saw, but we had a confrontation of sorts over our center panoramic lens. Tucker firmly vetoed our idea. He claimed that the horizontal spread of the lens on each side of center did not allow for the boat-like prow he had envisioned. He wanted the sculptured front to be cut back almost as far as the boat-shaped hood above until it joined the inside surfaces of the front fenders.

So, knowing that the client's wishes come first and foremost, we cut back on each side so far that the wooden buck underneath had to be excised to allow the penetration. This kind of surgery is not particularly unusual in clay studies, but it is messy. Wood chips and clay together mean trash, and we discarded globs of the residue. Then, we applied fresh clay and sculpted the new shape — and installed a sealed-beam headlight front and center, just as Preston wanted it. (My personal view was that the panoramic lens would have worked just as well, if not better.)

When Tucker returned to the scene, he confirmed that we were on the right track, but that the front end grille was still not quite, "that's it!" Back to the drawing board and clay tools.

Alex Tremulis and his two designers hired from Barnes and Reinicke, a local firm, worked steadily on the No. 1 clay model. They had refined it far beyond what we had first seen. The crew was doing a superb job of incorporating all the nuances of Tremulis artistry in their design. At the same time, there were aspects of the No. 1 model indicating a hearkening to a different drummer. Here and there were signs of Preston Tucker's acceptance of things we were doing on the No. 2 clay model. These accepted ideas were included on the Tremulis study.

This was quite understandable (even if a bit disconcerting to the Lippincott crew). Preston Tucker was running the show. He was the authority. It was impossible for Alex to be unaware of approbations accorded our efforts with regard to a hood shape or a roof contour. Also, design is a profession wherein acknowledgment of other designers' forward steps is a way of life; a designer on a desert island could not stay in the mainstream of ideas. Thus, Alex and his crew quite logically emulated some of the details of our work. After all, we had, from day one, emulated his basic shape of the body of the Tucker '48.

As the diary sketches indicate, the Lippincott crew were also devoting some degree of attention to details such as hood ornaments, tail lights, and trim, all of which are focal points of controversy in design far transcending their intrinsic value. Top management, in the role of client, will often accept a mediocre overall form (actually of vital importance) and bicker endlessly over the choice of a hood ornament design or the shape of a tail light lens (something which a future customer might not even be aware). One or two of us roughed out some hood ornament concepts on paper and sculpted them in clay. One of these designs was always sitting quietly on our clay model.

Tucker Madawick likes to tell the story about a Studebaker design by the resident Raymond Loewy Associates group, in which Loewy participated in a management session about hood ornaments. Loewy, with his persuasive manner, delightful French accent and grand demeanor, gave a convincing pitch for the need of a touch of elegance afforded by a splendid hood ornament. Standing before a completed clay model of an upcoming Studebaker, he took several Loewy staff model ornaments and placed them, one by one, on the car. Then he placed his final choice on the car — backwards. No one said a word. It was accepted as the design ornament for that car. "After all," says Tucker Madawick, "It probably lasted a year or two in production and then another design took its place." So it was backwards.

Tucker tail lights did not escape attention either. Our idea of a long bright metal form with a lens at the end, resting on top of each rear fender, became a *cause celebre.* A number of Tucker personages put in their two cents on how it could be achieved. It was really quite simple: use a chromium-plated die casting for the front part and acrylic plastic molded in clear red for the lens. It would require die casting and injection plastic molds costing probably eight thousand dollars at the time, a pittance compared to the tooling costs for an automobile.

However, we soon discovered a reluctance on the part of the company to undertake such a specialized custom artifact. Due to time and financial constraints, almost all of the parts for the Tucker '48 were going to be bought outside and brought into the plant to be assembled. Thus the solution was to seek an existing tail light design and plan on using it. A pre-war Dodge design was eventually used. The incident highlighted an important chink in the armor of the Tucker Corporation.

In spite of the vast size of the factory and all of the machinery and machine tools which had been housed there at one time, hardly anything but the roofs, walls and floors could be used in making an automobile. An awareness of this had, early on, begun to vex the management, the engineers, the accountants and the purchasing personnel. Thus, in general, fabricating in-house parts was discouraged. Preston Tucker had definite plans to use the engine fabricating machine tools, formerly employed to make Wright aircraft engines, to build his own Tucker engine. However, the Tucker engine was still in the development stage and no one knew when it would require the tooling set-ups to produce it.

We began to bring various bright work (those that would be done in chrome) components of the Tucker '48 to completion. The next step was refinement. No matter how skillfully sculpted, a muddy clay surface intended to be a bumper or a section of trim, does not adequately convey the impression of chrome. Special treatment is required.

Aluminum foil is applied to the surface piece by piece, then carefully formed against the clay by hand. Aluminum foil does not bend into compound shapes easily, so small pieces are used. Shellac is sometimes brushed on first to assure adhesion, but it often results in overflow if the shellac is not applied very sparingly. The area is then carefully burnished by hand or cotton to bring it to a luster. I believe that

Tucker Madawick was the only one of us who was a professional of this technique. He had done it many times at the Ford Motor Company, where, he said, "it was done so well that you couldn't tell our clay and aluminum bright work from the real chromium thing." With Tucker M. to guide us, we applied the foil to each new area as soon as it was ready for the treatment. (photos no. 13, 14).

By this time it was April, and the unpredictable Chicago weather inflicted a heat wave upon the non-air conditioned Tucker plant. Our infernal clay began to sag. Hundreds of pounds of the stuff, painstakingly ladled onto the buck, had begun to move and the completed surfaces refused to respond to polishing to a final smooth patina. We tried turpentine as a glossing agent; it did help, but the sagging continued. Then, in a moment of inspiration, one of us called the Tucker Corporation fire department. "Bring over your CO_2 truck quick!" we said. The firemen weren't particularly busy, except for occasional drills, and they probably enjoyed a genuine call, but the odd nature of the request did not escape them. Still, in rapid order, they brought over the truck and expended cylinder after cylinder of carbon dioxide on the models to cool the clay. CO_2 released into the atmosphere is a chilling fog and it did the trick, successfully curing the sags. From then on, whenever it was too warm in the development bay we sprayed our magic nostrum on any surface ready to be polished — or about to sag.

Even as we completed the clay work, many of our original sketches and renderings still hung on the walls in our studio and on our backdrop partition. We frequently referred back to them, sometimes adding new ideas, and other times removing old ones. One rendering always stayed, however. It was the one Read had done in New York of a front bumper with a dip in its center section, a sort of steer horn effect. Oddly enough, it just kept hanging on.

As in any team project, duties are relegated or just naturally fall into place, and specialization evolves. Hal Bergstrom was the organizer, Read Viemeister and Budd Steinhilber the more prominent sources of design inspiration, Tucker Madawick, the expert on handling clay, and I became the car's interiors and instruments man. We all contributed meaningful ideas to the styling of the car, so the final result incorporated something from each of us.

Tucker knew that he had to get the metal prototype completed in the shortest time possible in order to maintain the public's interest and to keep the money coming in. As Budd later said, "As the designers labored over their mock-ups, Preston Tucker continued to fight the dragon of financing the project." Tucker had set his sights on presenting a completed and working prototype Tucker '48 to the world in early summer. The completion of the sheetmetal form in the development bay was an urgent consideration.

Work on the metal prototype continued, but the tempo had gone from allegro to adagio since we had entered the scene, with smaller and smaller portions of finite shapes being given to Herman Ringling and his crew. As Tucker, his engineers and Alex all accepted one part or another of the no. 1 or no. 2 clay model as final, that particular shape was measured from the clay and translated into sheetmetal.

The exterior was proceeding in clay (and slowly in metal), but the interior had hardly been touched. To get this aspect underway, I was assigned the task of working out a make-do instrument/control design to be installed in the prototype. A number of my sketches and renderings of this area of the car had provoked interest, including the idea of putting all driver controls to the immediate left and right of the steering column. I was given a prewar Studebaker instrument cluster from a derelict car and told to incorporate this into a convincing representation of Preston Tucker's message, the Tucker '48. This was indeed a challenge. Although I revelled in the assignment, I knew very well that it was not first class. No one, I hoped, thought that this was going to be the production design (photo no. 24). We had the pattern shop assemble a rough interior mock-up of the driver area, a basic frame for my design of the prototype driver center. This included an idea for a steering wheel with a flat bottom for comfortably resting a hand (while cruising at 100 mph?).

Like many facets of the project, the master interior was going to be accomplished, "later." Little did I realize then that the instrument/control cluster would remain my responsibility right up to production design. This "yes, but later" syndrome began to worry several of us. Tucker Madawick was disturbed by the apparent lack of planning-in-place for volume production. He knew that ultimately the project could succeed only if thousands of cars issued from an assembly line every month. A "we'll let tomorrow take care of that" attitude seemed to prevail which focused on getting a prototype ready for presentation first, and working out the essential details later.

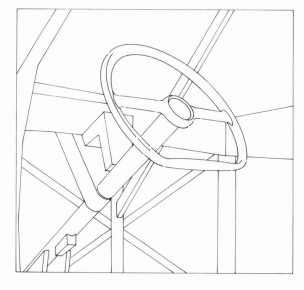

Activities in the development bay re-enforced this impression. While the two design crews still worked on the clay models, a crew from Barnes and Reinicke began to take critical measurements using the bridge and its metal rods. Something new was stirring, obviously brought on by an order from high authority.

The contrast of our environment to that of a typical major automobile styling studio at this stage was vivid. Normally, the entire clay model was complete and approved by top management before proceeding to the measurement phase. Every scrap of excess clay sliced off the model would have been scraped off the platform with military attention to housekeeping standards. In our case, we were still carving clay and sculpting some areas of the car not yet measured. Fortunately, the movement of the bridge back and forth was not an impediment (it is moved inch by inch), but we still trampled underfoot our cast-off clay. I could not help think of the apocryphal story of the gas station attendant, filling the tank of a huge gas-guzzling car and telling the owner of the vehicle to, "...turn off the engine, you're gaining on me!"

Actually, we *were* gaining. The fifth week on the project had arrived, and the no. 2 clay model was in excellent shape.

The rear fender air intake had been refined, (photo 15). The rear quarter window had been swept back to provide more visibility. Door handle ideas were now represented, and subtle changes to the trailing edge of the front fender had been suggested. The grille was a new treatment (photo 16). There was no trim on the catwalk (the area between the hood and the front wheels). Parking lights had been incorporated in a horizontal band on each side of center, and there was a new Tucker logotype (done by a graphics team at the New York Lippincott office) placed just below the Cyclops eye. The vertical bars in the grille had vanished as we began to move toward simpler solutions. Our latest hood ornament was in place, and we were in the processes of studying wheel disc ideas.

The rear of the car had evolved convincingly with the exhaust pipes *in* the aluminum foiled bumper (photo 14). The plinth upon which the name *"Tucker"* was affixed was at the center of the bumper. There was trim at the end of the roof sweep, but the rear window was too high. This was an unavoidable problem brought about by the need for adequate access to the rear engine.

The two clay models were almost complete. Model no. 1 was very close to the March 2 newspaper rendering except for the rear fender air intakes, which were still being worked out. The no. 2 model was just short of climax. Some of its trim was extraneous (and the corporation was loath to have it anyway), and the front grille still lacked the dramatic touch the Tucker '48 demanded. But both were very convincing Tuckers. Like two thoroughbreds racing toward the finish line, their moment of reckoning was drawing near, and at the moment, all that was missing was the roar of the crowd (photo no. 17).

The reasons for the sudden stir of activity in the plant were revealed to Hal Bergstrom on Monday, April 21 (day 41 for us). Tucker called a meeting of the fifteen key persons responsible for making the design a reality. Hal and Alex Tremulis were among those present. Tucker had decided to mobilize a crash program to assemble six cars, regardless of obstacles, with the first of them, "on wheels, painted, trimmed and out of here Saturday night. . . .The whole cockeyed nation is watching us," he said, adding, "We better stick to the fender on the number one clay model. I want that straight line on that fender."

It was as if a skipper had, alone, become aware of having passed the point of no return in a voyage. Why Preston Tucker chose this particular time to propose such a monumental undertaking has never been revealed. Perhaps he knew that it was impossible, but believed that his image of the impossible dream would spur all of us into achievements which he could not otherwise expect.

Incidentally, we never did find out *which* straight line on the fender he was referring to.

CHAPTER 9

CONSUMMATION

Watching a crew working on the bridge surrounding a clay model is fascinating — for a while. It is actually a laborious process; inserting the rod into the bridge until it just touches the model and calling out the measurements, repeated inch by inch, hour after hour. The man on the bridge must have a complete absence of acrophobia. It is no trivial assignment, being asked to climb the bridge and straddle it with rod in hand, eight feet over a tender clay model.

Watching this work being done on our models, reminded me (by my sweaty palms) of several days I had spent at the Columbia University engineering laboratory in New York while working for Edo Aircraft in 1943. We were stress testing to destruction, the main landing gear support panel for a giant aircraft float project using the large multi-story hydraulic press in the lab. The test panel was at the top floor level, with hydraulic mechanisms above and extending two stories below, through a large gaping hole in the floor. We had to walk a narrow plank out to the panel to install strain gauges. I quickly acquired a lasting respect for anyone who has to work in elevated surroundings.

As the home stretch came into view, the distinctly separate nature of the two clay models gradually changed. Everyone knew, particularly Preston Tucker, that there could only be one design for the Tucker '48, regardless of whether it was all no. 1 or all no. 2 or a combination of both. The two competing projects had become increasingly interdependent. It simply wasn't possible for each to ignore the travails of the

other. Alex didn't ignore what we were doing and made many suggestions which helped us. We, in turn, contributed a few ideas to him;
certainly a just aid to a worthy compatriot. A constrained, delicate and
very successful rear fender air intake on the No. 1 clay model was one
of the results of this cooperative effort. Ultimately, it was Alex's
accomplishment, but Read and Budd contributed ideas which helped
him carry it off (photo no. 15).

Deep in the psyches of all of us in the Lippincott crew was an urge
to go all out, but the constraints of our original charter from Preston
Tucker did not allow such a departure. Still, it seemed only logical
that, as long as we were in Chicago, we be given the opportunity to
depict the very best Tucker car we could offer. Why not, we thought,
while finishing the Lippincott model, convert one side of it to a post-
48 design, perhaps a Tucker '50 or '52?

Tucker did not object to the idea, and so we translated many of our
most avant-garde sketches into clay. At the same time, we were, still

A quick sketch of Read Veimeister's steerhorn bumper

concerned with the final 1948 front grille. Fine tuning was essential. We revived that long-dormant steer-horn bumper and carefully sculpted its graceful form into the front end of what the Lippincott team finally presented as the Tucker '48 — and the Tucker '52.

The great Raymond F. Loewy,* lived by the credo, "Never leave well enough alone." In the final stages of our seven week project at the Tucker Corporation, we were inspired by that motto, and spent every last minute going over clay and foil to make sure that everything was in the best possible order. A last look at every line, every curve, every detail became paramount (photo no. 18).

Tucker visited frequently, making many last-minute suggestions concerning contours and small details. The model's left side advanced design was accomplished, and at last, with its Viemeister bumper, the no. 2 model was whole. The result lacked one final detail: paint. No clay model, regardless of how well received, can ever be signed off until it acquires the smooth glossy luster of a realistic painted finish.

The acrylic windows, the laboriously aluminum foiled bumpers, the wheels, and other details were carefully masked. Everything was ready. The finishing department supervisor, whose task up to now was primarily planning the facilities for painting future Tucker '48s, was called upon to personally paint the two clay models. A neutral gray color was chosen to reduce color prejudice — "I don't like it! It's not red.", or "Why didn't you paint it blue?".

Outside, it was a warm April day. The development bay windows were opened to release the clouds of highly flammable lacquer spray that were about to be released for hours on end. Tanks of paint, compressor and hoses were at the ready. The supervisor, dressed in slacks and a jacket, stood poised, spray gun in hand. Welding torches were turned off, and other precautions were taken to assure that no sparks would be emitted in the area.

The supervisor lifted his spray gun and began to work. "My God!" one of us shouted, "You're smoking a cigarette!" A lit cigarette dangled in his lips. He looked up surprised, casually put it out, and

---

*Raymond Loewy died on July 14, 1986 at the age of 92, after making the term "industrial design" a part of the lexicon. The architect Philip Johnson once said Loewy, "started industrial design and the streamlining movement."

went on to do a splendid paint job on both models.

A masked car, real or clay, is surprisingly like the Hollywood film image of a person swathed in gauze after a plastic surgeon's magic, who upon removal of the bandages, is revealed in their new self.

Carefully, very carefully, the masking paper and tape were removed and bit by bit, the new glossy grey cars emerged as the lacquer dried. They were both very convincing Tucker '48s.

Presentation day was not heralded with trumpets and photographs. It just happened. Preston Tucker arrived to view the results of months of work: two finished full-sized models (actually representing three designs with the dual-faced Lippincott specimen) brought into being by his wish, his drive, and his confidence in the mission. He beamed. He walked around both models, obviously pleased. He took one look at our steer-horn front end and said, "That's it!" But he would not indicate further what he would chose as the final design, in whole or in part. We had a general idea of what portions of the no. 1 and no. 2 models he liked, but we had no clear cut composite in our minds of what he would determine by fiat to be the ultimate Tucker '48. I'm not even sure that he knew at that moment. He seemed to like the steer horn bumper, most of the front end and the rear grille of the Lippincott model. Our swept-back rear quarter window, rear fender air intake and small details such as our hood ornament were in limbo. The '52 left side was definitely "Not now, maybe later." (photo 19).

There was no fanfare, no champagne, no ceremony. Alex Tremulis had been working on the project for 125 days at least, we for 53. The consequences of what all of us had done now devolved upon Preston Tucker. He was very complimentary and obviously eager to have the metal prototype completed. No mention of that six day, six-car crash program was made, then, or to my knowledge, later. It was not six days, but six weeks before the finished prototype revealed what his design choices actually were.

The completion of the Lippincott model on May 3, 1947 was the end of outside participation in the design of the Tucker '48. The Lippincott crew, our job completed, dispersed. Read went back to Yellow Springs, Ohio, and the rest of us departed for New York. Our participation in the Tucker saga had taken two months of single-minded attention. It was a fascinating experience alloyed with the drama of uncertainty. We all felt that we had succeeded. Yet, as we departed, we did not know how much of our creation would see fruition in the ultimate Tucker '48.

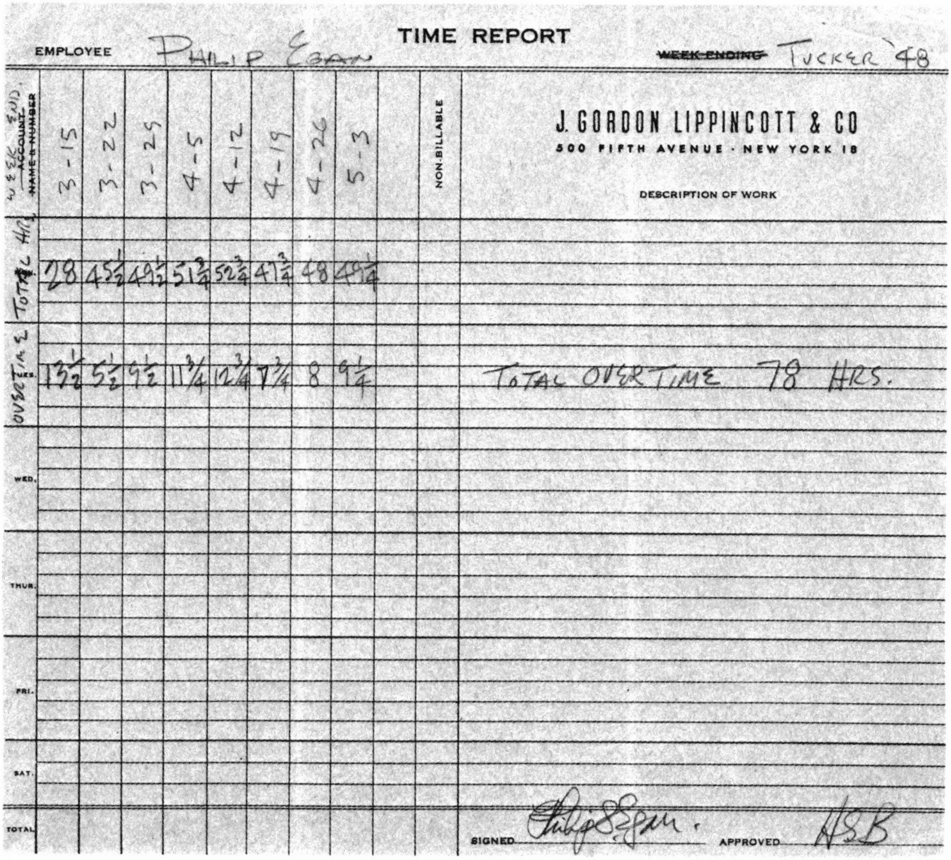

TIME REPORT

EMPLOYEE Philip Egan

~~WEEK ENDING~~ Tucker '48

J. GORDON LIPPINCOTT & CO
500 FIFTH AVENUE · NEW YORK 18

Week end. ~~Account Name & Number~~	3-15	3-22	3-29	4-5	4-12	4-19	4-26	5-3		NON-BILLABLE	DESCRIPTION OF WORK
Total Hrs.	28	45½	49½	51¾	52¾	47¾	48	49¼			
Overtime	13½	5½	9½	11¾	12¾	7¾	8	9¼			TOTAL OVERTIME 78 HRS.
WED.											
THUR.											
FRI.											
SAT.											
TOTAL											

SIGNED Philip Egan. APPROVED HSB

The author's time card shows the record of activity for Lippincott's Tucker '48 project.

PART II
DESTINY

CHAPTER 10

THE TIN GOOSE

When inanimate objects become important to people, they are often given nicknames that express affection or derision. Alex Tremulis gave such a sobriquet to the metal prototype Tucker '48 — *"The Tin Goose."* Although the nickname was later grossly misinterpreted by literal minds bent on discrediting the entire Tucker project, it was definitely an indication of affection.

Alex Tremulis was primarily responsible for guiding the fabrication of the Tin Goose to conclusion. He was privy to Preston Tucker's decisions regarding those portions of the no. 1 and no. 2 clay models that would be shown to the public. The logistics were mind-boggling. Alex had to coordinate his colleagues in sheet metal forming, body engineering, engine and drivetrain design, interior furnishings, instrumentation/controls and painting to produce the final product to the satisfaction of the boss. It had to be beautiful, it had to be convincing and it had to run.

The *run* consideration was the crux of the entire project. In the six weeks between "That's it!" and "Ladies and gentlemen, the Tucker '48!", a radical engine concept had to be assembled for the first time, and it had to run reasonably well. There was no time for major changes. In a discipline where engine designs normally require years to achieve success, weeks were being allotted.

Once the two clay models were finished and decisions made about what parts to copy, the bridge crew sent their measurements to Herman Ringling to translate into steel. Tucker Corporation body engineers then lofted sections of the design onto paper and onto sheet metal master patterns.

A clay model is a three dimensional artform. The people responsible for converting its curves and measurements to a hand-made prototype bolt and weld the artform together. They then smooth out the hammered, bolted shapes and their junctions by grinding and by filling the gaps and ripples with body solder. Then a flame-heated flat paddle is moved across the molten solder to blend it into the body shape, and the surface is ground smooth.

The body engineers wish to minimize the labor involved in making and fitting all of the automobile's frame and body panels. They study every form in relationship to those adjacent and plan where each part will adjoin the others. This detailed planning is based upon knowledge of how sheet metal can be pressed into shape and cut by dies to minimize the rework, welding, body soldering and grinding.

The creation of Herman Ringling and his crew emerged as a masterpiece of compromise with the incessant changes in contour, detail and direction. Budd Steinhilber recalls that he saw Herman cutting with "a hacksaw into a section of the front fender to make a change and there were seven layers of built-up sheet metal!" If ever a Cellini had a seemingly insurmountable challenge, this was it. Begun on a much-altered Oldsmobile frame, the Tin Goose evolved into a shining form which beautifully reflected earlier sketches, renderings, clay model measurements and verbal instructions.

Between the 3rd of May and the 19th of June, the personnel of the Tucker Corporation rallied to bring the product of Preston Tucker's obsession on stage in an unveiling comparable to the opening night of a Broadway extravaganza. This was a twenty-five million dollar production, and while perhaps not everyone in "the whole cockeyed nation" would be watching, there would be countless stock holders, dealers and automobile enthusiasts eagerly awaiting the opening curtain.

Now on the sidelines, the members of the Lippincott crew would also be watching to see what parts of their efforts had been built into the final revelation.

June 18, 1947 — the day before the unveiling of the prototype Tucker '48 — started off well enough. Countless invited guests, many of them signed-up dealers, gathered in Chicago to see this sensation of postwar cars. Tucker had a section of his huge factory roped off, bannered and festooned. White folding chairs striped the vast floor in endless rows. A large band had been hired and an immense stage set up, high above the floor, with a ramp to the left leading to the north plant courtyard. The stage was bordered with thousands of flowers, its backdrop a diaphanous drapery twelve feet high with an oval Tucker family crest on a four foot placard. On this stage the Tucker Tin Goose would debut, brought forth from her crucible for all to see running under her own power.

Back in the crucible the object of all of this attention was being readied for her grand entrance. She was beautiful, the Tin Goose. Every square inch of her was hand made, all of her sheet metal patterned bit by bit after the clay models in the development bay. It had been like hammering sheet metal from a live model. There it is in clay; that's it, that's the shape, make it! She was painted with a splendid gloss of rich maroon, the color chosen by Preston Tucker.

Everything had all been done to perfection, but it soon became apparent that Lady Luck would not bestow any special favors. On the eve of the unveiling day, the Tin Goose was hundreds of pounds overweight from welding and solder-filling and from twenty-five layers of paint. All of this weight was supported by wheel suspension arms designed for a normal Tucker '48 — the production version, still in the future.

As an engineer put the finishing touches to a part of the car, he slammed shut a door and the jar was too much. One of the suspension arms collapsed. Preston Tucker's dreams of a velvet ride included those arms. They were held in place by rubber bushings, instead of metal springs, to absorb road shock. With only hours to go before presentation, Tucker's dream was listing. Then, the distorted load became too much for another of the suspension arms and it too gave way.

The situation was perilous, but Lady Luck had not completely abandoned the maroon Pauline. Despite his enthusiasm for Tucker's engine, William Stampfli did not share Preston's faith in the suspension arms. He had disagreed, first openly and then secretly, with the metal specified for these critical members. "They should be of aluminum," Tucker had said. "Steel!" Stampfli contended. Aluminum they were, Tucker being boss. However, Stampfli had covertly ordered two or three sets, one of aluminum. the other of steel, to serve as spares (some accounts say that a third set, of manganese bronze was also ordered).

Tears streamed down the faces of the harried engineers and mechanics watching the sinking of the Tucker '48. "Never mind! We have the answer," said Stampfli springing into action. "Someone help me! I have replacement arms that will work. Come on!" The glistening Goose was jacked up and in fantastic time the steel arms were installed.

The sleepless night stretched into morning as last-minute preparations continued. The crowds arrived, some 5,000 strong, all impatient to see the car. The Goose at last seemed ready, but Stampfli's pessimism should have been more expansive. The huge Tucker 589 engine had finally been assembled, but testing had barely begun. The specimen engine in the prototype had been run only briefly. When started up it gave out a roar loud enough to raise the dead. Time was pressing hard now, the throng in the improvised auditorium was becoming restless; the show had to go on regardless of the noisy engine.

Tucker ordered the band to play its loudest and to keep the sound level up as the prototype was rolled up to the presentation area and mounted the stage behind the twelve-foot curtains. But the music could not cover a new problem. As the engine warmed up, its coolant flowed to the front-mounted radiators and started to boil over.

No one was in any mood to hesitate any longer. Steam or no steam, it was time to go ahead. Gene Haustein, the Tucker test driver, climbed in, and set the transmission lever to "go." The curtains parted, and the Tin Goose came onto center stage. No one noticed the problems. The Tucker '48 Tin Goose prototype was gorgeous. Haustein shut the engine down as soon as the car reached center stage, and the crowd, all 5,000, rose to its feet and roared its approval!

The men who had worked the Tin Goose into shape ran across the courtyard expecting to see the worst, and were delighted to see the best; their creation was a standing-ovation success.

Preston Thomas Tucker did not have to sell anyone anything that day. While neither a body engineer nor a stylist, he had somehow guided the project to a conclusion that no one could question. He had worked long and hard to get to this moment, and with thousands cheering his creation, Preston Tucker beamed. (photo no. 20).

For 1948, the car was distinctly dramatic. Gone were the bulbous, jutting contours of the typical postwar rehashes. Every line seemed to express a cooperation with the air rushing past, allowing smooth passage over the surface with a minimum of disturbance. It was long and low and wide and had not one jot of extraneous protrusion, nor one excess of trim. Its message was even austere. (photo no. 21).

The look of the car was a commanding vindication to those who styled her: to Alex Tremulis, whose body design had prevailed and succeeded, to the Lippincott crew, who designed the ultimate front and rear, and to the perseverance of Preston Tucker. The Tin Goose was also a tribute to Herman Ringling and his associates. It was they who hammered out the Tremulis body contours, and the Lippincott steer-horn bumper and transverse rear grille. The blend of Tremulis, Lippincott and Tucker was a huge success in the eyes of the cheering crowd.

A few days before the inauguration, still in the midst of exterior and interior refinements, the Tin Goose was rolled out into the daylight for a photo session. The car's chrome steer-horn bumper had not yet arrived (although the metal rear bumper had), so a painted wood mock-up was substituted for the cameras. Bright metal inlays had been added to the spaces in the front grille. The deeply sculpted contours on each side of the Cyclops light displayed Read's bumper quite dramatically. (photo no. 22).

The Tucker logotype, in spaced metal letters, was taken directly from the 1950 side of the Lippincott clay model and placed on the side of the Tin Goose. This placement of the name was later abandoned. Production-line Tuckers had the logo on the rear bumper. That sassy rear end had been achieved. There was no way anyone would ignore a Tucker '48 as it roared past. (photo no. 23).

Preston Tucker had every reason to be proud of his maroon delight. His choices in the car's size and appearance had panned out with golden results. However, the mechanical aspects — the hesitating response of the Tin Goose — were an ominous portent. The fluid drive couplings at each end of the engine were straight-through drive. They had no provision for reverse. The chorus of cheers was accompanied by a murmuring discord of this discrepancy — the car would not go in reverse. Regardless of any rejoinders by the Tucker Corporation that this was a temporary matter, the onus of a car that couldn't back up hung like a pall over the otherwise glad tidings.

CHAPTER 11

OUTSIDE AND IN AGAIN

The impact of the Tucker's unveiling before thousands was dramatic. Everything that could be said about the gleaming car was told to anyone who would listen. The glitches of its rather *"steamy"* presentation were generally overlooked as the price of experimental development. After all, this was the first hand-made vehicle; it was bound to have a fleck of clay here and there, literally and figuratively.

The impact was not confined to those who saw the car newly born. The staff of Lippincott & Margulies (the company's name was changed in the summer of 1947) eagerly absorbed every new article, photo and advertisement about the Tucker. The Lippincott crew of the no. 2 clay model had no regrets. The Tin Goose almost said it all.

Gordon Lippincott was among the ardent supporters of the Tucker. He was pleased to see the extensive full-page advertisements of the car in national newspapers and magazines. Preston Tucker had hired a New York public relations firm in the summer of 1947 to stir up enthusiasm. When the P.R. people learned that Gordon Lippincott's design staff had styled the Tucker '48, they called him. One of the top radio shows of the time featured "Dunninger, The Famous Mind Reader." Gordon recalls that, "The P.R. firm figured if Dunninger could read minds, perhaps he could also read drawings. Would I be willing to sketch a Tucker on the show?"

Gordon thought it a good idea, and at a "trial run" luncheon meeting, Dunninger startled everyone present with his ability to read quick sketches of trivial subjects done by Gordon, hidden from the master of extra sensory perception. "The publicity people were delighted," recalls Gordon. "We decided to go on the air. I was to do a picture on a big flip-over pad, four or five feet wide, in charcoal, of the Tucker automobile with its Cyclops eye and he (Dunninger) was to guess what I was drawing."

Gordon drew the picture, and Dunninger, off-stage, said, "He's drawing an automobile. It's only got one headlight."

Gordon told Dunninger to come out to see for himself. "Oh, it's a new automobile," he said. "What is it?"

It was the perfect lead-in. Gordon said, "It's a Tucker and it's going to be on the market in a few months."

Tucker's publicity people scored a bull's-eye that night. The need for enthusiastic word-of-mouth about the Tucker '48 had reached a nationwide audience. And Gordon Lippincott became a convert to real-time ESP.

Gordon was far from alone in his enthusiasm for the Tucker project. When I resumed my duties at Lippincott and Margulies, I established a telephone correspondence with Alex Tremulis to keep up to date on the Tucker project. He gave me a running commentary in his inimitable style. The Tucker project continued to fascinate me and we tentatively agreed I would become his assistant in Chicago if I left Lippincott. Product design at Lippincott and Margulies had slowed considerably, and as the junior member of the staff, I was politely discharged by Walter Margulies. In one of those grand coincidences, Alex called almost that same day to tell me that he had received approval from Preston Tucker to take me on as his assistant. I readily accepted, and in July 1947, I became a full-time staff member at the Tucker Corporation.

Alex Tremulis welcomed me as his assistant as if I were already a friend. He let me know from the first day of my employment at Tucker that my talents were highly regarded. While it was true that I had been in the midst of the New York styling world, I was actually a neophyte and regardless of my talents, I had much to learn. I took his regard as a great compliment.

He had established a modest but adequate styling studio off one of the hallways through which, only a few months before, I had walked on the way to the development bay for the first time. The echoes I had heard then were now much diminished. Almost all of the offices in the administration building were occupied. Across the hall, the newly enlisted staff metallurgist was busy setting his gear in place, including pilot plating tanks to establish quality control standards for incoming parts. Our thirty by ten foot design studio had a large east exposure window which overlooked a courtyard. During the next few months, that courtyard became the scene of energetic testing of the Tucker '48 production models.

The studio had drawing boards in abundance and space to spare. Alex was an engaging boss. I listened intently as he told me about all that had transpired in my absence, including a colorful description of the unveiling of the Tin Goose. He also revealed that immediately after the unveiling, Preston Tucker ordered the total destruction of both full-sized clay models. They were hacked to pieces and thrown into the trash. No one ever learned why Tucker wanted this done. I shuddered at such a waste of time, effort and materials.

Alex outlined my initial assignment — designing the driver control area of the car. The one in the Tin Goose had been ineptly patched together for the prototype showing (photo no. 24). The final design had to be created from scratch, starting with sketches and renderings for Preston Tucker's approval. The specifications for the interior basically stemmed from Tucker himself.

In a preliminary meeting with Alex and me, he demonstrated his zealous concern for safety. His concepts of a safe interior included: a padded crash pad across the top of the dash; all driver controls clustered around the steering wheel area; omission of a panel forward of the front seat passengers to provide a large empty crash pit into which — according to Tucker — a passenger, seeing an imminent collision, could seek safety. Seat belts had occurred to Tucker, but he felt that they would imply something inherently unsafe about the car, that the Tucker '48 was too vigorous, too fast for anyone's good.

Tucker did not specify an exact design or style, but he was intrigued by my idea of aircraft throttle controls for lights, choke, throttle, etc. It was the beginning of a very pleasant and exciting period of automobile design for me. I had already been quite sure that Alex Tremulis would be an equable boss. And my first meeting with Preston Tucker gave no hints of difficulties in the future.

Decisions on the Tucker's interior design were actually much overdue. Elsewhere in the plant, the body engineers were intent upon completing the lofting of full-sized contours gleaned from the two clay models so that dies could be made to shape the steel of the car's body parts. However, nothing substantial had been done with the car's interior. Alex and I began layouts for a full scale interior mock-up.

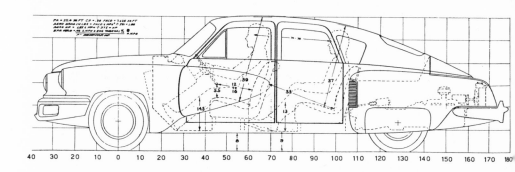

In the Tucker pattern department we found an angel, a model making genius in Terry Moffatt. Terry was a contemplative, capable soul who had the ability to understand what a designer had in mind almost immediately. A basic sketch, a brief explanation of a line here and there and Terry comprehended the idea. In a few days (or sometimes even hours), he came back with the idea carved in wood for confirmation, and then, shortly after, presented a beautifully crafted finished rendition. He also had the ingenuity to use the skills of other artisans in the plant to add the touches of paint and polish to provide us with a finished presentation-quality model. His level of expertise was that of top-drawer industrial design.

Immediately next to the design studio was a small unclaimed office which promptly became the interior mock-up studio. Here, part by part, the interior of the Tucker began to take shape in wood, sheet metal, seating, flooring, pedals and a steering column. Everything was arranged to represent the real car. All of the essential dimensions: headroom, seating height, distance to the windshield, had to be accurate. Our mock-up had to be suitable for the incorporation of the finished model of my driver control area. At times, Terry was ahead of us as we worked on our layouts. "O.K. what's next?" he would say. I don't think that Terry wanted to do plain pattern work; the more nearly complete concept delighted him.

My sketches evolved into presentation renderings for Preston Tucker, and I recall that on the third try they were accepted as "That's it!" Then it was on to the drawing board to generate layouts and model drawings for Terry. In little more than a week he began installing the parts in the mock-up, and within two weeks Tucker approved of what had been done thus far.

One detail was not yet approved — the steering wheel. I had been quietly working on a design which I hoped would satisfy Tucker's passion for safety, and still be a design first. It would have a large flat energy-absorbing hub with a generous horn button in the center. The horn was adorned with the Tucker family crest in injection-molded acrylic. The horn ring, proven to be lethal in collisions, was eliminated. Alex liked my full-sized pastel rendering of the wheel, and we had Terry Moffat make a model of it. The finishing department painted it ivory beige. We showed it to Tucker. I was certain he would love it.

Tucker looked carefully at the design for a few moments and quietly said, "I know that what I'm going to say is not strictly along safety lines, but what I really want is something like my Cadillac's steering wheel. It has a small pointed hub; small enough for my hand to grasp. On an icy road, I can feel the ice under the front wheels, and control the car with my hand on the hub." Thus one learns of the idiosyncrasies of one's client. My energy-absorbing steering wheel was out. Subsequently, a design by Schiller (a prominent steering wheel manufacturer) was accepted. The hub would feature my design of the Tucker family crest. The finished product never made it into production, although samples of the hub were run, (photo no. 25).

Thus, minus the steering wheel, detailed drawings of the entire driver control area became my top priority. I drew the extended oval instrument panel as a large injection-molded acrylic panel surrounding a circular speedometer with status instruments within the circle. Just to the left of this area were the driver control knobs, while to the right, the radio dial and knobs barely protruded into the passenger area. Outside sales engineering experts were consulted over each aspect. For the acrylic panel (which was huge by the standards of the time), I consulted the

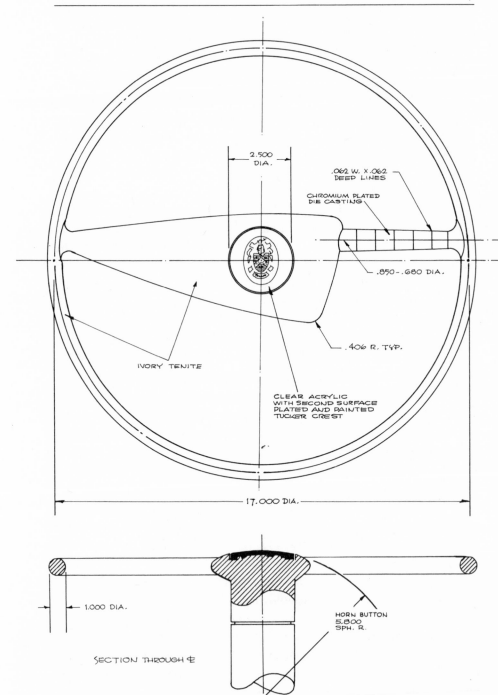

2.500
DIA.

.062 W. X .062
DEEP LINES

CHROMIUM PLATED
DIE CASTING

.850 - .680 DIA.

.406 R. TYP.

IVORY TENITE

CLEAR ACRYLIC
WITH SECOND SURFACE
PLATED AND PAINTED
TUCKER CREST

17.000 DIA.

1.000 DIA.

HORN BUTTON
5.800
SPH. R.

SECTION THROUGH ₵

Chicago representative of Rohm and Haas, a prominent manufacturer of acrylic compounds.

Stewart Warner, a Chicago company specializing in instruments, sent in their sales engineer. He suggested using an existing instrument cluster that could be modified to any graphics I wanted. Something I wanted was a twelve o'clock speedometer. This showed zero and 120 m.p.h. at twelve o'clock, with divisions at watch dial increments. Stewart Warner's engineer said, "No problem. We'll simply set the pointer to start where you want it and use a movement which goes 360 degrees for 120 m.p.h. The graphics will be yours."

The Tucker radio was entrusted to Motorola, a pioneer in car radios. Bob Galvin, who later became Motorola's chairman, was then their Chicago headquarters sales engineer. He suggested we use an existing Motorola radio model which could be turned on its side and modified to our own specifications.

By fall, the many pieces of the Tucker project were gradually falling into place. Many of my original concepts, tempered by reality, were being vindicated. Alex worked on concepts for a suitable hood ornament, seating and interior refinements. He also drew up ideas for future models of the Tucker line of automobiles, including a convertible. We both worked on a great variety of side-projects, including graphics for Tucker dealer showroom signs, which we hoped would soon be seen all over the country. We were also assigned design work on the Tucker premiums, those little incentives offered to dealers for promotion of the car. One premium was a Tucker ashtray designed around the form of the Indianapolis race track (I thought it was a low point in design).

Almost every Tucker employee knew that time was precious, every moment had to be fully utilized, yet mundane details had to be considered. One of our styling department projects was the Cyclops eye. This central beam in the nose of the Tucker '48 seemed, at first glance, to be a marvelous simple idea. Each twist of the steering wheel, would illuminate the dark area to the left or right of a turn. Fine on paper. Two factors clouded the brilliance of this concept. Alex explained that tests had revealed that the beam of that single headlamp did not adequately penetrate the two main headlamp beams. Discussions of using a panoramic lens in future models to overcome the deficiency brought back vivid memories of the aborted Lippincott efforts to design that very thing, for that very reason.

The other problem was that the 48 individual states did not have uniform codes regarding acceptable automobile lighting systems. Each state had a myriad of laws governing details of automotive engineering. Fifteen of these states had specific requirements for *two* headlamps, (plus or minus zero) for the front of a car. Every one of these states would have to be laboriously lobbied to overturn these laws — a lengthy process at best.

Alex and I assumed that the legal answer in those states would be to advise Tucker owners to disconnect the center headlight. What would happen to a Tucker owner crossing into a non-complying state was never resolved. The alternative finally chosen was to design a decorative cover to place over the Cyclops eye of any Tucker '48 sold within these state boundaries. This project was given to us, although it had very low priority.

Preston Tucker's Cyclops eye serves as a singular example of the countless concepts he thought up which were just that — concepts. Even ideas that worked well, such as Tucker's pop-out windshield, which would automatically break loose when hit with sufficient force (a force less than fatal to a person's head), had unforeseen consequences. Some employees found that the windshield would just as easily pop out by placing a wet plumber's plunger on the glass and pulling on it. A car thief who knew this would have an easy time breaking into a Tucker.

The concept is the genesis, be it a self-starter to avoid hand cranking to start the engine, or the fully automatic shift. These and numerous other ideas start with the need and require countless hours, and often years of research and development to bring to fruition. Tucker didn't have this time factor, but in his gauntlet thrown down to the automotive world he included concomitant challenges to his engineering staff. "Here! I've thought of this; create it!" The Cyclops eye was one of these.

William D. Perfield of the Tucker engineering staff came up with the linkage and fittings that made the Cyclops eye work. The central headlamp was on a pivot, connected through a rod and fulcrum system to the left front wheel. Switches kept the center headlamp *off* when the car steered straight ahead, and turned the light *on* when the car was turned ten degrees or more to the left or right.

Every decision took us closer to the edge of that inevitable cliff — the final result. That last plunge would have to answer all the questions: does it work, do the parts fit, does it really look as good as proposed? Normally the final result is not a matter of waiting days or weeks, but rather, months, or years. Dies and molds to shape metal and fabricate plastics are complicated and expensive. The engineer and designer put their necks out with every significant component or product created. A mistake finds them in a very lonely setting. The scenario at the Tucker Corporation was a definite cliff-hanger.

CHAPTER 12

THE BIG ENGINE THAT COULDN'T

It was now well into the third quarter of 1947, and feverish activity centered around another focal point of the whole picture — the engine and its drive train. Several chassis had been rapidly built to contain prototype engines and transmissions for road testing. At frequent intervals Alex and I were roused from our drawing boards by the thunderous roars from one of these creations on the ramp outside our studio.

A raw powered chassis is quite a sight: a frame with wheels on suspension arms, a steering wheel, improvised seats for a driver and recording assistant, the engine and transmission, battery (or batteries, in this case), a fuel tank and all manner of wires and pipes. All of this essential gear is completely exposed to the outside world without a body. Gene Haustein, our chief test driver, treated the test chassis with intensive care. It worked, to a point, but the Tucker Corporation grapevine was circulating ominous rumors of severe side effects. At first, we took these as being the usual teething problems of a new engine design. All of us knew that the Tucker project badly needed to have a proven powerhouse of an engine.

While the chassis buck was being tested day in and day out, a part of the powerplant question was still unanswered, how to cool it. In December 1947, Alex learned of a small wind tunnel facility at the Armour Institute of Technology.* We asked Terry Moffat to build a

---

*Armour was the precursor of the Illinois Institute of Technology, made famous by the great architect, Mies Van Der Rohe.

precisely contoured, one-eighth-size wood model of the Tucker '48. The result was beautiful. It was the first accurate representation any of us had seen of a true Tucker. This was the Tin Goose, clay models number one and two, and all the changes to date, presented in miniature — the real thing divided by eight.

Alex and I tenderly put this model in the hands of the Armour wind tunnel engineering staff for analysis. They were to put the miniature Tucker '48 through an extensive series of tests to determine its aerodynamics. We sought a reading of the flow of air, the pressures and antipressures, the smoothness or turbulence, and the general behavior of the airstream past the shape of the car. Particularly, we wanted to know what happened at the rear where that elegant grille had been placed. The Tin Goose had been temporarily fitted with radiators behind the front grille. Their steaming, bubbly performance on presentation day showed that a rear-engined car dictated cooling means adjacent to the source of power. We needed to know whether the air should be sucked in at the rear or pushed out.

Armour did its work. The results showed us that we needed to bring the air in under the body and through the rear fender air intakes into the engine compartment, then past the engine and through a fan or fans to a radiator located against the inside of the rear grille. This knowledge at last provided the answer to a major question of the Tucker '48 project. Armour also told us that the drag coefficient, the measurement of the Tucker '48's streamlining, was remarkably advanced for its time. In fact, contemporary automobiles do not significantly better this drag coefficient.

While the wind tunnel tests answered the engine-cooling hypothesis, elsewhere in the Tucker Corporation, alarm bells were about to clang. Months of testing the huge Tucker engine with its fluid couplings on the sparse chassis buck seemed to have reached an impasse. After numerous talks with the engineering staff, Alex brought back ever-gloomier reports of the results of Gene Haustein's unflagging efforts. I recall one time when Alex told me that dynamometer testing (putting the engine on a test stand and measuring its output) had indicated power as low as eighty-three horsepower, less than half of its anticipated rating. (photo no. 26).

A bank of batteries was necessary to deliver enough power to start the engine. The hydraulic valve system did not work until the engine had been cranked endlessly, and most of the heat generated while running the engine went out the exhaust pipes. The fuel injection was a no-show. On and on the reports told a story of failure.

A seasoned test driver, Gene Haustein, found this lack of a breakthrough frustrating. Alex told me that one day, Haustein informed Preston Tucker that "the engine has acceleration like the moon coming up, and it sounds like a barrel full of monkeys with the lid propped open!"

This certainly wasn't the first criticism Tucker heard about the 589 engine, but it must have been pivotal. Shortly after, he made it clear that a new engine must be found — now. The frantic search that ensued also involved the Tucker's transmission. When an automobile's engine is placed over the driving wheels, the resulting vehicle is inherently different from the average front engine, rear drive car. Because the space for power and its delivery are in the same area, the engine and transmission are virtually a unit.

Going to a different engine meant possibly abandoning the one key advantage of placing the engine directly over the rear axle centerline — the almost ideal weight distribution (promoted extensively in Tucker advertisements). Suddenly, Preston T.'s idea of a double-ender engine with power going directly to the rear wheels was *hors de combat*, simply because there were no such engines available. Consequently, the new engine would most likely be located behind a transmission placed on the rear wheels' centerline. This configuration had presented problems in the early Czechoslovakian Tatras with all of that engine weight hanging out to the rear. (In recent years, Tatra has changed to an engine location over the rear wheels axle centerline in their 613 Special.) Whether Tucker knew of this characteristic of the Tatra I do not know, but certainly key staff members of the Tucker '48 project were aware of the potential for problems.

By the late winter of 1947-48, the delays in following through on the glowing success of the Tin Goose presentation reached a crisis in the minds of stockholders, dealers and the eager public. The failure of the 589 engine was not a secret, nor was a succession of postponements with other segments of the project. The travails of fund raising, the scratching of fluid drive to the wheels, abandoning fuel injection and giving up on a transverse engine position, all put the reality of the

Tucker '48 in question. That promised first car off the production line was nowhere in sight. The quagmire into which Preston Tucker was falling threatened the very existence of the Tucker Corporation.

CHAPTER 13

THE SMALLER ENGINE
THAT COULD

Serendipity intervened. The result was one of the most interesting episodes in automotive history.

The H.H. Franklin Manufacturing Company of Syracuse, New York, began manufacturing air cooled automobiles in 1902, rode the crest of the 1920s boom, and then fell to oblivion in the Depression year of 1933. Carl Doman and Ed Marks, two of Franklin's key engineers, survived the blackout by manufacturing and marketing Franklin air cooled engines for truck and industrial use under the name Air Cooled Motors, Inc. The name "Franklin" was never officially used after the H.H. Franklin Co. folded, but because of its excellent reputation, it was freely used to describe the engines. Doman and Marks had mastered the nuances of engineering air cooled engines for automobile use. Though controversial, and not readily accepted by the general car-driving public, these air cooled engines were power packages complete with cooling fan and shrouds for installation in light trucks and vans. During World War II they quickly penetrated the aircraft market, where air cooled engines were the standard, and in great demand.

Their enterprise endured through WWII and became the property of Republic Aircraft Corporation. Republic invested over three million dollars in Air Cooled Motors and immense sums in their post WWII amphibious plane, the Seabee. The SeaBee was powered by a 215 h.p. Franklin air cooled engine. (The interior of this plane had been designed by J. Gordon Lippincott & Company.) Unfortunately, the

SeaBee did not prove to be a big seller, and by 1947 Republic was in dire straits.

Republic's subsidiary, Air Cooled Motors, was not without other resources, however. The company had achieved considerable success in the aircraft engine market, including Bell helicopters. One of the engines for Bell was the 6 ALV-335.

According to an account by Carl Doman,* the Tucker Corporation first became aware of the Franklin 335 engine when Max Garavito, president of the Tucker Export Corporation, came to Syracuse one Sunday in July 1947, and met with Doman and other principals of Air Cooled Motors. He expressed great interest in using the engine in the Tucker car, if it could be watercooled. Doman and Carl Roth, head of Air Cooled, then went to Chicago to meet with Preston Tucker and sell him on the new engine. Tucker listened intently, and then said, "It is just too bad that we didn't get together earlier, but now our engine plans are definitely set. We are going to build this six-cylinder engine with Ben Parson's hydraulic valve mechanism."

One of Tucker's reasons was that the five inch (127mm) bore of the big engine was suitable to the capabilities of the machine tools which had been used in making Wright aircraft engines at the plant. These machines were just sitting idle, and Tucker wanted to use them. Doman and Roth went back to Syracuse to concentrate on other markets for their engines.

Even at this early date, when the presentation cheers for the Tin Goose had barely faded, there were those in the Tucker Corporation who perceived terminal problems with the 589 engine. Nonetheless, testing of the big engine continued into the fall and winter months, with no success. Finally, at the end of 1947, Preston Tucker made the decision to scuttle the 589 engine, and hired Paul G. Wellenkamp as the new engineering manager of the Tucker Corporation.

Doman relates, "Mr. Wallenkamp came to Syracuse to discuss engines. He asked us to quote on a group of twenty-five liquid cooled engines, with proper flywheel housings and accessory equipment. We agreed to submit a quotation in ten days.

*Carl Doman's, "Tales From The Business of Life" was serialized in "Tucker Topics" the newsletter of the Tucker Automobile Club of America, editor Richard E. Jones. Vol. 5 no. 2 through Vol. 5 no. 7, 1978.

"In the course of this conversation, Mr. Wallenkamp was very frank and told us of his transmission problems. The torque convertors which Mr. Tucker had discussed so optimistically just didn't work. Then, there were no proper production transmissions which could be obtained. Fred Loetterle, our chief draftsman, said, '. . .why don't you use the Cord front wheel drive transmission, but mount it in the rear.' The idea seemed worthy of exploration, so we went down to the Pare Garage in nearby Liverpool and borrowed a Cord repair manual. Fred made a sketch to show his approach to the installation. It was so good that Wallenkamp said he would try to work it out when he got back to Chicago."

The quote on the twenty-five engines came to a grand total of $125,000, including forging dies for the crank-shaft, patterns for the block, cylinder head, intake manifolds, exhaust manifolds, flywheel and flywheel housings. Carl Doman's experience with automotive engines was clearly evident in the expert way he and his staff planned to convert the aircooled helicopter engine to a liquid cooled automobile power-plant. For example, flywheels are not found in aircraft engines, but are routine for all automotive engines.

"But," reports Doman, "the Tucker Corporation reported back that the price was way out of line. We learned by the gossip route that Tucker had made a deal with Jacobs Engine Co. to use its opposed six cylinder engine converted to liquid cooling."

Jacobs was a competing helicopter engine company, strictly an aircraft engine outfit, albeit one with credentials comparable to Franklin. Doman and Air Cooled Motors seemed to be out of the running. The Tucker Corporation desperately needed an engine and it looked as if a Jacobs engine would be it.

Then Doman relates a new twist in the Tucker story, "One day in early February of 1948, Frank Chadwick, our Parts Sales Manager, mentioned that the Ypsilanti Machine & Tool Company had ordered many miscellaneous parts for a 6 ALV-335 engine."

Frank Chadwick soon learned that this machine shop in Ypsilanti, Michigan, was owned by Preston Tucker's mother, and that the company had recently taken delivery on three complete Air Cooled 6 ALV-335 engine assemblies purchased from the Bell Helicopter Company. On hand to receive the goods were Eddie Offutt, the chief experimental engineer of the Tucker Corporation, Dan Leabu, and Preston Tucker's twenty-two year old son, Preston Jr. These men spearheaded the effort to convert the 6 ALV-335 engines to liquid cooled automobile powerplants, and pair them with a renovated 1936 Cord 810 transmission. Doman says,"Ed Offutt, without drawings, had designed liquid cooled cylinder blocks and cylinder heads, manifolds, flywheel and flywheel housing. And as a result, he had a fine operating power plant." (photo no. 27).

The engine was built of seven aluminum castings, two cylinder heads, two blocks, two halves of the split crank, and an oil pan. Offutt, Leabu, and Tucker Jr., replaced the aircraft style finned cylinders with steel jacketed hollow housings to allow flow of coolant around each cylinder. With a displacement of 335 cubic inches (5489 cubic centimeters), a bore of 4.5 inches (14mm), a stroke of 3.5 in. (89mm), and a compression ratio of 7:1, the new engine weighed 320 pounds and delivered 166 horsepower with 372 foot-pounds of torque. The torque was so great that, from a standing start, it was possible to break the teeth of the first speed gears of the Cord 810 transmission. Forty years later, prestige performance cars have not matched this rated torque. For example, in cars with higher horsepower: Cadillac Allante's 170 h.p. to Mercedes Benz 560 SEC's 238 h.p., the torque only ranges from 230 to 287 foot pounds.

Another race against time developed, and this time it was no trial run. Preston Tucker had to show the real thing to stockholders and dealers in the upcoming first annual Tucker Corp. Stockholders' Meeting, March 9th. This time the car could not be a prototype Tin Goose, but an actual Tucker '48 with its finished production body and efficient engine. It had to look good and run well, with that vigor and pizazz Tucker spoke of so often.

The Ypsilanti engineering trio had just finished one of the new engines. The night before the stockholder's meeting, Ed Offutt brought it from Michigan to the Chicago plant and installed it in a pilot run '48. The next day at the meeting, "Preston Tucker announced that at 11:00 a.m., he would demonstrate the car," says Carl Doman. "At 10:55 a.m., the car was ready and Preston Tucker demonstrated the car by driving it across the Dodge Chicago plant at a speed well above 60 m.p.h. Quite naturally, a cheer went up from the crowd. At last they were convinced that the Tucker Corporation was about to produce cars. I might add that Mr. Tucker, to dramatize the demonstration, requested that the mufflers be left off the vehicle. Yes, it was also a very noisy demonstration.''

Carl Doman's Air Cooled engine and a twelve year old transmission carried the day. The beautiful car ran with great vigor and as a bonus, it could also go in reverse: both firsts.

CHAPTER 14

THE TUCKER '48
IN GEAR

The Tucker '48 was at last becoming a reality. Tucker body parts, ordered from eight major Detroit fabricators of dies for the stamping and trimming of fifty-two body parts, began to arrive at the plant. Some of the metal body parts for the Tucker were unusually excellent in spite of their being made on Kirksite dies. These are dies cast of a relatively low temperature metal, rather than machined from hard steel. Kirksite dies are used to save cost and time, and are usually resorted to for relatively low volume production runs. It is presumed that Tucker would have eventually replaced these with steel dies for his planned high volume production runs.

At last it became possible to put portions of the Tucker '48 together and to organize a pilot production line (photo no. 28). To the delight of the growing numbers of visitors, it was finally beginning to happen. The car was real. They could actually see Tucker bodies in white (parts assembled before painting) moving along an assembly line. By early March, 1948, Tucker '48 serial number 1001 (number one), 1002, 1003, and others began to take shape. Throughout the entire plant, morale was the highest it had ever been. There was open admiration and respect for the engineers and technicians who had made the translation of the design one of quality. It was good, superbly good. This judgment was to remain valid through all of the days of pilot production (and continues to this day in existing Tucker '48's).

Among the innumerable routine and special parts that were shipped in were fifty steering wheels from the Ford Motor Company. After Preston Tucker rejected my asymmetrical proposal, he chose a steering wheel by Schiller, but precious weeks elapsed before Schiller was given the contract. They immediately began tooling for the new design, but could not deliver the wheels in time to be installed in the pilot production Tuckers rolling off the assembly line. The problem nearly caused panic until Alex Tremulis came to the rescue.

Alex had many contacts in the automobile industry and called a friend at the Ford Motor Company. Upon hearing the problem, the friend responded, "Alex, we'll ship you fifty Lincoln Zephyr steering wheels we have in stock. They have slight blemishes, but you can use them. All we ask is that, when you get your production wheels, you remove ours, destroy them and replace them with your new ones."

"O.K. great!" Alex said. "Send us an invoice for the cost."

"Forget it," Alex's friend replied, "they're yours. Just destroy them when your new ones come in."

The pairing of the splendid pilot production bodies with a suitable engine created a genuine entity for the first time in the history of the project. Success, as Winston Churchill said, is having failure after failure with confidence. Now, the testing of the actual car could begin — on a limited basis at first because of a shortage of engines and transmissions.

For the time being, Cord 810 and 812 transmissions had to be found and used with the engines. This was meant only to be a stop-gap measure until new Tucker transmissions could be produced. Dan Leabu, Preston Tucker Jr. and others scoured the country's junk yards, and twenty-two Cord transmissions were found in from good-to-poor condition.

The Cord transmissions were all re-engineered at the facilities in Ypsilanti to adapt them to the Air Cooled Motors engine, and to overcome inherent faults in their design (gear tooth characteristics made them inadequate for the high torque to which they would be subjected). Producing these vital Ypsilanti versions of the Cord transmission slowed down production in Chicago to some degree.

The Ypsilanti-Cord transmissions were always thought of as only stop-gaps. A contemporary automatic transmission — "automatic" in 1948 terms — had to be developed for the long run. Warren Rice, a brilliant young engineer, had come to the Tucker Corporation in 1947 to head a transmission development program. Rice was following Carl H. Scheuerman, Jr., who had been hired in February of that same year to engineer the fluid coupling transmission for each end of the 589 engine. After the demise of the entire transverse engine-transmission idea, Scheuerman worked on other transmission projects until he was assigned to redesign the Cord transmission for production at Ypsilanti. The Scheuerman manual transmission was labeled the Y-1 (Ypsilanti-1).

Warren Rice was given orders to develop a truly automatic transmission for the Tucker '48, '49, and the years beyond. This was essential for survival in the automotive market. An American family car could no longer be offered with just a manual stick-shift transmission. Rice and his team developed the R-1 Tuckermatic transmission. One was installed on a pilot assembly line Tucker, and another placed in a test chassis, both with good results.

From early spring of 1948, when the first cars were assembled (the first twelve Tuckers have re-worked Cord transmissions), until the delivery of the initial Ypsilanti (Y-1) and Rice (R-1) transmissions, these two programs ran in parallel. This resulted in a race which proved to be very interesting to watch from the sidelines.

As Preston Tucker was candid enough to admit, the Ypsilanti prototype engine required more redesign for production. The original quotation from Air Cooled Motors of $125,00 for twenty-five engines, had been rejected by the Tucker Corporation earlier in the year as being too high. However, when the Ypsilanti project of converting the 6 ALV-335 helicopter engine worked, the Air Cooled engines were, "That's it!" Now, that engine had to be designed for final production.

There was only one source of supply and redesign of the powerplants, Air Cooled. When their chief engineer, Carl Doman rebuffed an offer from Preston Tucker to become his director of engineering, Tucker responded by asking, "Is Air Cooled Motors for sale?" Doman replied, "I think it is. Let's call the president, Carl Roth." Roth met with Preston Tucker in Chicago, talked the matter over and the upshot was that Tucker flew to Syracuse to inspect the plant there. The result was not an immediate offer of purchase, but an assurance of engineering contracts. Tucker obviously had more than that in mind, because a week later, he and his executive vice-president, Fred Rockelman, agreed to fly to New York to attend a meeting of Republic Aviation's board of directors — a board anxious to divest Air Cooled (their SeaBee venture no longer needing Franklin engines in profusion). Before the day was over, Tucker privately asked Carl Doman, "How much should I offer?" Doman said, "I would suggest $1,800,000." Tucker responded with an offer of that figure and the Republic board "reluctantly" accepted the offer.

With two checks totalling $1,800,000, Air Cooled Motors of Syracuse, New York, became a subsidiary of the Tucker Corporation of Chicago. Preston Tucker made the announcement of this purchase on March 21, 1948.

By March, Alex and I felt that the basic industrial design of the interior was on target — the crash pad, the new look of the instrument and control area, and the general impact of the layout (photo no. 29). However, the driver controls were only part of the interior. The styling, color and fabric choices of the interior were still rather pedestrian. We needed an expert in such details. Several months before, Alex had begun to search for a fabrics and color stylist to round out our design crew. At last he struck gold with Audrey Moore, an interiors stylist who had been with Studebaker in South Bend, Indiana. Audrey joined our twosome in mid-March and rapidly demonstrated her ability to size up the situation and politely but expertly point us in the right direction. With excellent sketches and renderings, she brought reassurance that the production models of the Tucker '48 — and beyond — would be top drawer (photos no. 30, 31).

The May issue of *Tucker Topics*, a monthly company publication, then in its sixth issue, carried an article which featured the modest Tucker styling department chaired by Alex Tremulis. The two-page photo filled article, entitled "They Gave The Tucker '48 That New Look," formally introduced our team to everyone at the Tucker Corporation.

One of the photos in the article showed me working on a ⅛ size clay model of a possible future Tucker design, (photo no. 32). This was a continuation of work first done by the Lippincott team — sculpting the left side of our full scale clay model into a 1950's Tucker. Preston Tucker named this two-door model the "Talisman," and hoped it would be a second line of cars produced by the Tucker Corp. (photo no. 37).

At Tucker's behest, Alex and I also spent considerable time developing concepts of a small car. The plan was to use an Italian chassis of about the size of what was later regarded as a compact car. We were given layouts of a projected Savoia-Marchetti (one of the great names in Italian aircraft design) chassis as a foundation, and used these for dozens of sketches and renderings. It was an interesting design exercise, but nothing came of it.

Another of Preston Tucker's challenges to the automotive world was his wish to use disc brakes on the Tucker '48. Tucker wanted to move these devices, reputed to be the ultimate for high performance automotive vehicles, from the race track to the highway. Alex and I believed that the spirited road work which a Tucker '48 would often experience would generate a great deal of heat in the discs. Some obvious aesthetic and functional treatment seemed appropriate to emphasize their presence and to lower the temperature inside the wheels. We proposed that the Tucker's wheel discs be fitted with scoops to push air past the disc brakes. They would look good and provide needed cooling drafts. Our wheel disc design, later fitted to pilot production cars, was prototyped and installed on a 1948 Hudson which the corporation had bought to analyze and to use as a chase car for road testing the Tuckers.

The Hudson was an excellent automobile. With a long wheelbase, step-down floors, a 60″ height (matching the Tucker), huge interior, and superb handling characteristics, it was the closest thing to a Tucker '48 ever brought out by another automobile company. With Marshall Teague at the wheel, the Hudson became a prima stock car in the early 50s with countless racetrack wins. I bought one in the winter of '48/49. It was like owning a land-borne speedboat.

We set the Hudson up on jacks in the plant garage to test the efficiency of the "turbine" scoop design. The wheel discs were fitted to the rear wheels. A driver started the engine and slowly brought it up to various speeds, indicated by the speedometer connected to the drive train. At 30 mph, Alex lit a cigarette and blew smoke at the whirling wheel disc. The smoke circled the wheel — nothing much else.

At 45 mph, the swirl was more energetic, but not very convincing. At sixty, there was a modicum of passage through the turbine buckets, but certainly no blast of air. The scoops looked good, but a little head-scratching revealed that the wheels of an automobile, even at 60 mph, may appear to be racing, but they are actually only going at about 600-800 rpm, hardly enough for a turbine, which generally functions most effectively at many thousands of revolutions per minute, not hundreds. Furthermore, we knew that the discs would have to be left and right handed for each side of the car to provide the same scooping movement of air — not very practical.

So much for functional air scoops on wheel discs, unless the vehicle is at Indy and running at genuinely high speeds. Some contemporary cars now have left and right hand "turbine" wheel discs. These cars will probably not be run at Indy speeds very often, however.

In one of my now-regular visits to the assembly area, my timing was perfect! A car had just been completed, oiled, and fueled up, and was ready to be taken out for its test run. A driver moved in behind the wheel, and started the engine — it gave a most dramatic sound through its six exhaust pipes. The driver eased the car through a doorway and into one of the still-empty cavernous buildings of the plant. At the end of a wall of windows, about two hundred feet away, was the large exit door. Completely enthralled, I stood and watched as the driver pressed the accelerator and raced through the building, the light from the windows dancing off the glistening body as if it were a polished diamond. Just short of the door, the driver braked and those Dodge tail lights glowed back in brilliant red. In an instant, the car disappeared to the right and out into the daylight.

I wasn't the only person watching. The launching of each new Tucker '48 was an event and I am sure that those witnessing this with me were as encouraged as I was (photos no. 33-36).

Those were halcyon days at the Tucker Corporation. Somehow, in spite of insuperable odds, the dream was taking shape. You could kick the tires of a real Tucker '48, and early road tests revealed just how good a vehicle it was. It handled well. It was agile. The thrust of the Air Cooled Motors engine pressed the driver hard against the seat back and its low slung configuration gave an impression of great stability. In spite of all of the obstacles, it worked like a charm.

As salutary as the compliments were, they were all subjective. We wanted the car to be a success, and every piece of evidence which fed our desires was greeted with alacrity. Fortunately, objective testing was taking place, resulting in irrefutable findings and hard performance figures. In spite of minor glitches, these results were encouraging.

In addition, outsiders who test drove the Tucker '48 were glowing in their praise of the vehicle. Herbert D. Wilson, the automobile editor of the *Chicago Herald-American*, wrote in the May 2, 1948 issue of the newspaper, that his test drive indicated that ". . .the car loafs along at 80 with the throttle half open. . .the acceleration is terrific, extreme roominess, has excellent vision and a feeling of safety and solidity." Tom McCahill, automotive editor for *Mechanix Illustrated* magazine, was given his choice of several Tucker '48s to test drive. After choosing one at random, he reported in the August 1948 issue that the Tucker '48 was one of the ". . .greatest performing passenger automobiles ever built on this side of the Atlantic." He went on to say, "The car accelerated from a dead stand-still to sixty miles per hour in ten seconds. I opened the throttle on a straight stretch of highway, and was soon doing 105 miles per hour. This was the quickest 105 miles per hour I have ever reached."

There were more glad tidings to come, but the cheerful scene at the Tucker dream car headquarters was about to confront some hard facts and figures which had nothing to do with miles per hour.

CHAPTER 15

BELOW THE BELT

It seemed that the good news was so pervasive that the bad news was, at last, going to be vanquished forever. The Ypsilanti-based crew delivered the first production design transmissions based on the Cord model in June of 1948. These were, of course, based on a known design, twelve years old, whose inadequacies were understood and corrected. This would presumably be the stick shift option for Tucker buyers, at least for the early phases of production.

More remarkable, Warren Rice's program of developing an automatic transmission came to fruition at almost exactly the same time. His R-1 Tuckermatic had been designed and engineered by Rice and his team in record time, with three units fabricated under the supervision of the ubiquitous Mr. Stampfli. Two of these were put into action immediately, with one installed on Tucker '48 serial number 1026, the twenty-sixth car completed. Another was placed in a test chassis and put to rigorous testing. The results were immensely gratifying, in spite of teething problems inherent in any complicated new mechanism. The R-1 showed every indication of being the answer — or very close. The R-1 was to be succeeded by two revisions, the R-2 and R-3, the latter generally considered to the be the ultimate design — on paper.

The new engine subsidiary, Air Cooled Motors, was getting up to speed in delivering engines to Chicago on the original order for one hundred and twenty-five units. In fact, the target for car body production of "five Tucker '48s per day by 15 July," announced by the Tucker Corporation in a press release of April 28, 1948, was significantly short of engine deliveries. Tucker Engines, air freighted from Syracuse to meet the Air Cooled agreement, were gaining on car production.

This increasing number of engines arriving at the plant served to focus attention on some very controversial questions. When it was a separate company, Air Cooled's per-engine cost for the original order of one hundred and twenty-five engines came to $5,000. Now, owned by the Tucker Corporation, the tooling costs built into the original quotation could be considered as part of the costs of either entity, but the costs were still there. At best, creative accounting could be structured to separate the tooling costs and tag the engines coming into the Chicago plant at $1,500 each, with the Air Cooled subsidiary somehow absorbing the remaining dollars. The rub was that $1,500 was the anticipated cost of building an entire Tucker '48, based on a retail price of $2,500 at that time. Plans were in place to start on a redesign program to lower the production cost of the engine to about $125 each — in the future. In the meantime, the Tucker Corporation would be producing Tucker '48s at a loss.

To make a gross profit of, say, 12% of manufactured cost, the car would have to go out the door at only about two or three hundred dollars more than the then-current cost of the engine. Studies done by two independent research firms later confirmed that the cost of the car minus engine would probably be in the range of $1,075 to $1,441.

There was also strong disagreement on the ability of the Ypsilanti Machine and Tool Company to produce the Cord style Y-1 transmissions in volume enough to satisfy the stick shift option. Herbert Morley, a vice president of the Tucker Corporation, said later, "I busted my elbow pounding it around when I was arguing with (Preston) Tucker. Tucker knew perfectly well that the Ypsilanti Machine and Tool Company didn't have adequate equipment to build transmissions. I told him, '...you'll rue the day!'."

Further, there was no clear evidence of funded plans to build the Rice automatic transmissions. Tooling and production for such a project would take months as well as vast outlays of cash. Much of this could be considered speculative, but one detail was more immediately disturbing — only fifty sets of body stampings had been procured for production start-up. This was a miniscule quantity for an automobile company, and led to the question: did it have something to do with the ability to pay? Daily, the rumors began to mount that a basic flaw had emerged in the operation of the Tucker Corporation. Not in the car, not in its engine, nor its transmission designs, not in the acceptance of the car by dealers and the public, and not, essentially in Preston Tucker's dream for his car, but in the bottom line, money.

Nonetheless, most of us felt that this could be surmounted, somehow. The world's greatest salesman had raised over twenty-five million dollars so far; surely he could raise enough to pay for engines, bodies, transmissions, parts, salaries and wages to see the project over the last great hurdle.

Little did we realize that the last great hurdle would assume mammoth proportions, far greater than the simple question of money running low. The Securities and Exchange Commission of the United States was about to deal a body blow.

The Securities (stocks and bonds) and Exchange (selling of same) Commission was and is a most august and worthy body in overseeing what had once been a multi-ringed circus in the buying and selling of stocks and bonds in the United States. The stock market crash of 1929 occurred during a get-rich-quick frenzy of speculative buying (often on credit) and then selling as a stock issue climbed in value. After the spectre of the market crash and the Depression shocked the country, the SEC was given power to control stock transactions with a firm hand.

For reasons which must be left to another treatise, this same SEC set its sights on the Tucker Corporation. Those halcyon days, which had come at last to the Tucker Project in spite of tremendous financial problems, probably could have been sustained if not for the SEC.

On June 3, 1948, James D. Coolidge, the Tucker Corporation counsel was summoned to the SEC office in Chicago and informed that an investigation of the company was underway. He was told that the books and records of the corporation were required and that the investigation would be kept secret if these books and records were voluntarily presented to the commission.

Then on June 6, 1948, the illustrious news columnist and radio commentator, Drew Pearson, leaked a tip that an investigation by the

SEC "would blow the new Chicago auto firm higher than a kite." The next day, a Monday, Tucker stock that was selling for five dollars a share, dipped alarmingly. Preston Tucker's reaction to the news was to fight. The secret investigation was no longer a secret and he felt that the firm could not operate efficiently without its records. The SEC secured a federal court order for the records, which Tucker could not fight and he was forced to relinquish all Tucker Corporation records dating back to April 1947. Besides all the company books, those records included all engineering reports, many boxes of material.

Many of us began to harken to the countless rumors of impending disaster — the possible end of the entire project and of our jobs — but we hoped against hope that some knight in shining armor would make all well and right again. Such was not to be the case. One morning in June, I came to work as usual, showed my badge to the guard at the plant gate, walked to the design studio and was greeted by a disconsolate Alex Tremulis who confirmed an edict from the chief executive officer of the Tucker Corporation, Tucker himself, that the plant was closed.

Audrey Moore was there also and we listened to what little Alex knew of the details. He was going to stay on, at least for a while (probably at no salary), but we had no choice but to pack up and leave. We gathered up our meager personal gear. I assembled bits of evidence confirming my role in helping to design the dream car of the future, evidence essential for that portfolio a designer needs to land another position, and sadly departed.

It was not the last time Audrey or I would see Alex Tremulis. He was a supportive friend in the months that followed. We have kept in touch with each other ever since. However, it was the end of the active design effort of the Tucker '48.

What followed for Preston Tucker was one of the more glaring examples of misunderstanding ever occurring in jurisprudence. If it had not been so tragic, it would have been ludicrous.

Preston Tucker's response to the court order for the company's records was not only to close the plant. He also published a full page open letter in newspapers across the country charging that, "...most of the political pressure and investigations we have had to face these last two years can be traced back to one influential individual, who is out to get Tucker." Many people felt that they knew who the individual was, but many also thought that there was surely more than one man who was on Tucker's case.

Michigan Senator Homer Ferguson (R), has been regarded by some as the culprit behind the scenes. Perhaps he overreacted on behalf of what he regarded as his constituency — the Detroit automakers. There is no direct evidence that the Big Three — Ford, Chrysler, and General Motors conspired to insure the Tucker Corporation's demise. The SEC was the primary instrument beyond any doubt.

In early 1949, United States attorney Otto Kerner announced that the February grand jury would investigate Preston Tucker and the affairs of his corporation. The upshot of this pre-trial trial was the announcement that Tucker and seven of his associates were indicted on thirty-one counts of mail fraud, violations of SEC regulations and conspiracy to defraud. Co-defendants included Fred Rockelman, a vice president, Mitchell W. Dulian, general sales manager, Otis Radford, comptroller, Cliff Knoble, advertising manager and others, including Floyd Cerf, the head of the brokerage firm which managed the sale of Tucker stock.

The trial, which lasted from October 5, 1949 to January 22, 1950 (including one mistrial), was an absolute washout for the prosecution and resulted in the acquittal of all defendants on all charges. The defense rested its case rather than rebut the prosecution. Preston Tucker's defense attorney, William T. Kirby, stated, "It is impossible to present a defense when there has been no offense."

Countless zealots on the government side of the vendetta had abysmally misunderstood the behavior of a man and his followers pursuing a dream. Instead, they accepted the rancor of those who had disagreed with Tucker as being evidence of fraud, of dishonesty, and of deliberate intent to deceive. The jury agreed with the defense and Preston Thomas Tucker was delivered from a judgment designed to ruin him and his associates.

Even though Tucker was acquitted of all charges, his dream of a Tucker '48 was ruined. Nearly two years had passed from the time I had left the Tucker Corporation. In late July 1948, the plant had reopened. Tucker hired some 300 employees, mostly assembly line workers, and together the small but loyal staff tried to continue to

build and road-test Tuckers. However, the damage by the press, and
stock losses, was too severe, and by March 3, 1949, the corporation
was in receivership, and its assets (mostly consisting of Air Cooled
Motors which survived until 1975) were in the hands of the courts.

Tucker continued to dream of cars, and went to Brazil in 1951 to
develop a modular sports car he called the *"Carioca."* Not physically
well at this point, Tucker soon returned to the United States with the
Carioca still on the drawing board. He died in Ypsilanti on December
26, 1956. The death certificate stated "pneumonia", but Alex Tremulis
had a better diagnosis, "broken heart."

28. By March, 1948, production model Tucker '48's were rolling off this modest assembly line. The cars were real and company morale was at an all-time high.

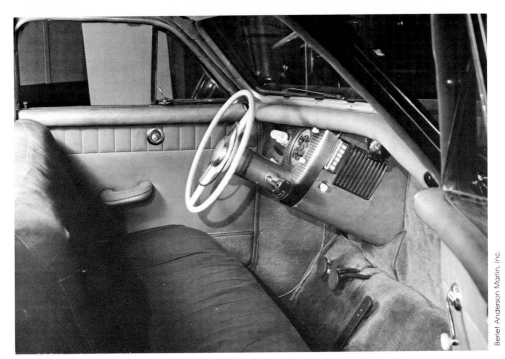

29. The production model instrument console, while still concise, was a vast improvement over the prototype's. The Lincoln Zephyr steering wheels were a gift from the Ford Motor Company.

30. The interior of the Tucker '48 was spacious and comfortable. The crashpad went all around the sides and front, and the arm rests were also padded. The "safety" door release button is from the Lincoln Zephyr.

Audrey Moore Hodges

31. Audrey Moore, interior stylist for the Tucker Corp., checking color chips to determine suitable combinations for Tucker upholstery.

John C. Cermak collection

32. Author Philip Egan is seen working on a ⅛ size clay model of a proposed two-door Tucker to represent a second line of cars.

33-34. These angled views of a production model Tucker, just driven off the assembly line, show the excellent lofting of lines by the body engineers. The smooth, flowing contours give the car an elegance rarely seen.

35. Viewed head-on, the tapered hood, three headlights, and gleaming bumper blend seamlessly together to form a dramatic visage.

36. Seen from behind, the Tucker '48 achieved that sassy rear end demanded by Preston Tucker.

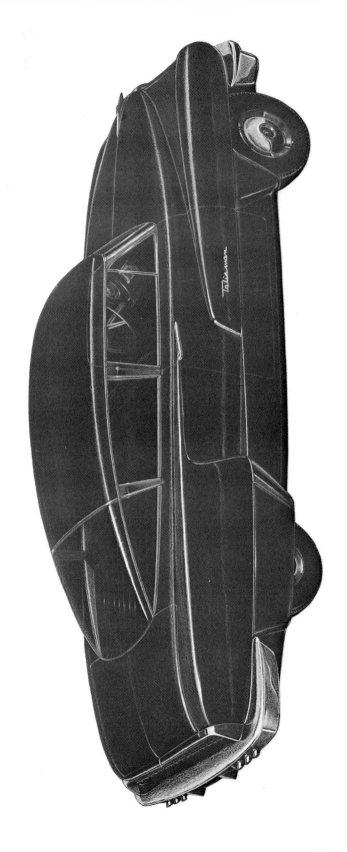

Philip S. Egan

37. This design, originally sculpted into the left side of the Lippincott clay model, was taken even further into the future by the author. This rendering of a 1950's two-door fastback was given the name "Talisman" by Preston Tucker.

38-39. The prototype "Tin Goose" is pictured here (above) with Tucker model number 41 (below), a fully restored car owned by Mr. Bev Ferreira of San Francisco, California.

The car always conveyed charisma.

40. Preston Thomas Tucker — "He believed in his goals implicitly."

Model T Ford

A Brief History of Automobile Design of the 1930's

Duesenberg SJ

41. 1930 Cord L-29 (United States).

42. 1933 Pierce Arrow Silver Arrow (United States).

43. 1935 Tatra 77a (Czechoslovakia).

44. The Tatra 2-603. This 1963-67 version is owned by D.E Barnett of Las Vegas, Nevada, and is probably the only example of this make in the United States.

45. 1933 Fuller Dymaxion (United States).

46. 1933 Briggs Tjaarda Sterkenberg (United States).

47. 1934 Chrysler Airflow (United States).

48. 1934 Citroen 7A (France).

49. 1936 Cord 810 (United States).

50. 1936 Dubonnet (Dolphin) Streamliner (France).

51. 1936 Lincoln Zephyr (United States).

52. 1938 Cadillac Sixty Special (United States).

53. Edsel Ford's first Continental. The 1939 Lincoln Continental was a distinctly refreshing design. E.T. Gregorie took the relatively advanced design of the Lincoln Zephyr into a new realm of styling with the Continental.

54. The 1938 Jaguar SS 100 — classic styling that is still copied today in replicas for proud new owners.

EPILOGUE

Thirty-seven Tucker '48s had been completed by the time the plant officially closed in the summer of 1948. Volunteer workers and later aficionados ultimately assembled another fourteen. All but three or four have survived to date and many are in running condition. Some of these are show specimens in museums. Quite a few are often driven by their owners and several have logged a quarter of a million miles on their odometers. Many of these are shown in automobile club rallies and win prizes regularly. Other Tuckers became derelicts and had to be restored or almost reconstructed. There are problems in keeping them going. Parts are available largely through channels dredged by fellow enthusiasts, and modifications have been extensive. Twenty years after the plant closed, I assisted Mr. Bev Ferreira of San Francisco, California, with the restoration of Tucker No. 1041 with some driver control parts husbanded through the years — just waiting for a Ferreira to come along. This forty-first car is the proud possession of an experienced driver and restorer of numerous classic cars who counts it as one of his prizes (photo no. 39).

Four decades after it went out of pilot production, the Tucker is an unqualified success with all of its owners, even though the car did not fulfill all of its vaunted goals. The engine is not directly over the rear wheel centerline, it does not have direct fluid drive, or disc brakes (only the Tin Goose was so equipped), or a sealed cooling system, or fuel injection, or driver controls limited to throttle and brake, or a speedometer on the hood, or a 24 volt electrical system, or an electronic ignition, and its selling price was never realistically established. (photo no. 38).

The absence of many of its promised features was a consequence of the state of the art in 1948. For example, disc brakes, fuel injection and electronic ignition were all still in the early development stage for automotive use.

The Tucker does have a superb engine in the rear (albeit aft of the rear axle), can cruise at 100 mph (160 kph), gives excellent fuel mileage (thirty miles per gallon at a steady thirty miles per hour in tests), has independent suspension on all four wheels, provides light steering, has a safety interior with step-down floors, the windshield will pop out on crash impact, has futuristic styling and is a delight to its few owners.

The car also evokes charisma — that special quality of leadership imparted by a man and his followers. It was perfect and yet flawed, it was a dream come true and yet unrealized. The Tucker Torpedo and then the Tucker '48 came at a time when people cherished dreams and seized upon those things which gave hope after the carnage of World War II.

Because the automobile is so crucial to man's mobility, it commands great attention. In this regard the Tucker '48 was a genuine leader. No other automobile in the immediate post-war era was nearly as dramatic. Although the production car did not abide to the letter of its proclaimed specifics, the final result was better than advertised and its problems could have been overcome. If allowed to go forward into full production, the Tucker would probably have become a maverick make joining the ranks of many notable automobiles. It would have continued to be different, innovative, and upsetting to its competitors.

The few surviving finished Tucker '48s are a lasting tribute to Preston Tucker, and to the many engineers and technicians who went him one better and made them work so well. Preston Tucker was a captivating leader and like all who succeed in establishing a loyal following, his winning attributes overwhelmed his flecks of clay. Eventually, forces beyond his powers vanquished him.

It is often claimed that there was a conspiracy against the Tucker Corporation mounted by Detroit, Inc. This conspiracy was supposedly a concerted effort by the existing automobile manufacturers and their suppliers to frustrate and delay the fortunes of Tucker by stalling the delivery of components. I have never found even one documented example of such behavior. Actually, the delivery of body stampings from eight die and forming sources in Detroit appeared to have taken place in a remarkably short period of time, and the quality of those parts was excellent. Tucker suppliers Rohm and Haas, Stewart-Warner, and Motorola — all suppliers to Detroit — delivered without significant hitches. One significant hitch, the lack of steering wheels for the pilot run cars, was due to in-house Tucker Corporation delay in decision-making. The Lincoln Zephyr steering wheels, given freely by Ford, were an open gesture of help by one of the Big Three for a newcomer.

The detractors of Preston Tucker, on the other hand, claimed he wasted most of the twenty-five million raised through personal and corporate extravagance and projects such as the 589 engine and other false starts. This complaint is not realistic. No new venture with the level of complexity to develop an automobile can be expected to be free of mistakes. The experiences of another entrepreneur serve to show what was needed — vastly greater funding.

Henry J. Kaiser, the shipbuilder who ventured into the automotive industry with Joseph W. Frazer to produce the Kaiser and Frazer cars, started with fifty-two million dollars in capitalization, and monumental industrial experience. In the first year of production, 1946, 12,000 very conventional Kaiser-Frazer cars were built and sold in competition with almost universally conventional cars of established makes. Production soared upward for a while, but as an ever-increasing variety of new cars (with engines and transmissions to match) came on the market, K-F experienced sales resistance. Even with a less conventional body style a few years later, sales plummeted and by the mid-fifties, production ceased. One reason for the major drop in sales was the lack of a state-of-the-art engine to match competition. K-F cars limped along with a six cylinder L head industrial engine, a variety used in taxicabs. Kaiser later admitted that he should have raised two hundred million dollars to finance his enterprise. With more start-up capital, he could have developed a suitable engine in time to sustain his earlier momentum.

Preston Tucker never had the momentum. His actual accomplish-

ments are all the more remarkable considering the capital he raised to
the day the plant folded was less than half of what Kaiser had found to
be inadequate. Time is money. Tucker didn't have enough money to
match the time required to bring his dream to reality.

Gordon Lippincott sums it up succinctly, "Preston Tucker
was honest. He deeply believed that he could produce this
automobile. I think that he believed that he could get it out
when he said he would. . .believed that it would be a better
automobile than anything Detroit had produced yet. He was an
honest, creative human being who simply took on too big a
project and under-estimated the time in which it could be
done."

The question of the Tucker '48's probable success in the hands of
the motoring public, will alas, never be answered. The Tucker would
have been the largest rear-engined passenger car ever produced. Unfortunately, the mass of the engine aft of the rear axle duplicated the pre-
war Tatra in configuration and problems. The condition was corrected
in later models of the Czechoslovakian vehicle, but Tucker's engineers
never had the time nor opportunity. Even in the relatively light-weight
Volkswagen Beetle, an engine aft of the rear axle proved to be less than
ideal. No one knows what problems might have occurred with the
Tucker '48 if thousands of owners had eventually driven them on icy
or rain-slick roads.

Potential problems and hypothetical corrections aside, enthusiasm
for the Tucker '48 has a sustaining chord in the Tucker Automobile
Club of America. Founded in September of 1972 by Richard E. Jones
of Orange Park, Florida, the group has grown to over two hundred
participants through the intervening years. Their annual Meet always
includes a number of running Tucker '48s proudly displayed. Members
include owners and many more non-owners whose motto is "Keeping
the Legend Alive."

APPENDIX I

THE GOLDEN AGE

A Brief History of Automotive Design of the 1930's

The history of the automobile to the beginning of World War II is permeated with ingenious approaches, superb inventions, weird ideas, and a high mortality rate. Through the processes of evolutionary development, rather than revolutionary change, the surviving brands were a relatively satisfactory combination of features and quality. Among the innumerable makes of cars world-wide, it is difficult to name a genuinely sensational volume production car; it is also difficult to name a thoroughly poor automobile which endured on the market.

To highlight this history, countless worthy efforts are going to be overlooked, but some basic criteria can be established to achieve a valid overview. One is engineering principles: how the basic car is conceived, where its engine is located and what means are used to generate power, together with essential layout, where the passengers are located and how engine or motor power is transmitted to the wheels. Another is how it appears within the envelope of its design concept; be it boxy or graceful, utilitarian or styled. These considerations, blended together, make an automobile, whether it is a Ford Model T. or a Duesenberg SJ.

In terms of the survival of any particular example of automobile, there is at least one other consideration — suitability to purpose. All of these ultimately become judgments by the final authority, the buying public.

Unfortunately for many great automotive inventors, designers and

engineers, numerous examples of vehicles appeared to meet these qualifications, but did not achieve or maintain acceptance. There is strong
evidence that revolutionary ideas do not succeed in the original, and
that many of the developments that do prevail do so under other
ensigns.

Among the following sampling of novel automotive engineering
principles, basic layout innovations and styling advances, are a few
which did not catch on, and others which did succeed.

1930 Cord L-29 (United States) (photo no. 41)
8 Cylinder Lycoming 299 cu. in. (4900 cc) engine
E. L. Cord, financier; Al Leamy, designer; Harry Miller, Leo
Goossen and Charles W. Van Ranst, front wheel drive advocates;
crafted the design of this aesthetic vehicle. Mechanically, the 1930
Cord left much to be desired, but its low profile, made possible by
front wheel drive and its details of styling came together to produce a
striking automobile. Defeated by the Depression and functional problems, Cord production endured only until 1932. All told, about 5,000
Cord L-29s were built.

1933 Pierce Arrow Silver Arrow (United States) (photo no. 42)
V-12 cylinder 462 cu.in. (7570 cc) 175 hp engine
One of the more prescient body designs in automotive history, this
flowing design had its front fenders (with the usual Pierce Arrow headlights in the fenders) swept back to form the body sidewalls which
were then punctuated by aircraft style rear fenders. Pierce Arrow stated
that "It gives you in 1933 the car of 1940." From a design by Philip
Wright, five Silver Arrows were built on 139 inch (353 cm) Pierce
Arrow chassis in four months at the Studebaker plant in Indiana. It
had one blemish; an uncomely rear window which was not designed by
Wright. The Silver Arrow had a top speed of 115 mph (155 kph).

1933-38 Tatra 77, 77a, 87 (Czechoslovakia) (photos no. 43, 44)
V-8 cylinder 207 cu. in. (3400 cc) 70 hp engine
A genuine revolutionary production car, the Tatra 77 of 1933 was
an anomaly for many years. Although the concept survived WWII
and has been produced since then, the elegance of this design begs the

question, why did it not have more obvious success? The answer may rest in the fact that it represented a focal point in automobile history, a point beyond which other designers and companies achieved great fame in building other versions of this grand idea. The design stemmed from the work of Paul Jaray, a key advocate of designing aerodynamics into automobile bodies. Jaray, an Austro-Hungarian, studied in Prague at the turn of the century, and early in his career was chief designer for the Zeppelin Works in Friedrichshafen, Germany. In the mid-twenties, Jaray shifted toward the application of dirigible shapes to auto bodies. Then based in Switzerland, he designed bodies for numerous automobile manufacturers in Germany and elsewhere. The 1920's and 30's were a pivotal period in automotive design in central Europe. In 1932, Jaray submitted a project design to the Ringhoffer-Tatra Works in Koprivnice, Czechoslovakia. The chief designer of Tatra, Hans Ledwinka, immediately employed Jaray's rear engine aerodynamic limousine to fit on the Tatra 57 chassis. Edmund Rumpler, Hans Nibel, Ferdinand Porsche and Hans Ledwinka (all Austro-Hungarians) were instrumental in bringing ideas of swing axles, central frame, torsion bar suspension, opposed cylinder air cooled rear engines and aerodynamic bodies from paper to reality. Tatra began to produce these cars in 1933.

Improvements were soon made in suspension and weight distribution of this full-sized rear engine car. If the abundant weight behind the rear wheels was not taken into account, over-steering could land car and driver in a ditch. This overhanging rear engine and swing axle proved less troublesome in the smaller scale Volkswagen Beetle. The famous Beetle was first introduced as the prototype VW3 by Ferdinand Porsche in 1936 Germany. So fundamental were the 1933 Tatra design and engineering features, that in 1967, Volkswagen of Germany paid Tatra over three million dollars for the use of Ledwinka's ideas and patents.

The Tatra was exemplary in terms of a logical blending of aerodynamic form with sound chassis concepts. It seated six, had an air-cooled engine and could do better than 90 mph (145 kph). By 1935, design refinements resulted in the remarkable 77a Tatra (photo no. 43). The Tatra was produced with gradual refinements until WWII and then after the war. There was no basic change in design from the

1930's to the 1960's; there did not need to be.

1933 Fuller Dymaxion (United States) (photo no.45)
V-8 cylinder Ford 225 cu. in. (3695 cc) 90 hp engine.

The brainchild of famed R. Buckminster Fuller, (father of the geodesic dome), the design of this 1933 vehicle is aerodynamically superior to most contemporary cars. Constructed more like an aircraft than an automobile, it rode on three wheels with a single directional wheel at the rear, propelled by the two front wheels through a differential axle. The Dymaxion's body was composed of aluminum panels on metal stringers over a chrome-moly chassis. Its shape was almost a true tear drop, or *"tropfen"* (liquid drop) in the classic German idiom so actively promoted by Rumpler and Nibel in the 20's and the 30's. Designed by Fuller and Starling Burgess, three were built. The Dymaxion car had a top speed of about 120 mph (192 kph).

1933 Briggs Tjaarda Sterkenberg (United States) (photo no. 46)
V-8 cylinder Ford 225 cu. in. (3695 cc) 85 hp engine.

John Sterkenberg Tjaarda, one of the most prolific visionaries of automotive history, designed this radical look into the future in 1933 for Briggs Manufacturing. The project was aimed at the Ford Motor Company's Depression-impelled need for a small Lincoln, a car between the Ford and luxury Lincoln automobile lines. A lineal descendent of Tjaarda's Sterkenberg streamliners, it was shown at the Chicago World's Fair in 1934. Built on a 125 inch (293 cm) wheelbase and weighing 2500 pounds (1133 kg), its Ford V-8 engine was mounted forward of the rear axle centerline with a semi-automatic transmission and differential coupled as a unit, easily removable for servicing or replacement. There were only two controls on the floor, the accelerator pedal, which was used to trigger the shifting of gears, and the brake pedal. A genuine mid-engine car, it had a top speed of 110 mph (176 kph). Only one car was ever built, but it was the progenitor of the Lincoln Zepher line of Ford Motor Company cars in years to come.

1934 Chrysler Airflow (United States) (photo no. 47)
8 cylinder 298 cu. in. (4881 cc) engine.

Promoted extensively, the Airflow (also produced as a De Soto with minor differences) was heralded by Chrysler as being the aerodynamic shape of the future. However, its tub-like fenders both front and rear, and lack of flowing lines did not show well to the public eye. With a unit body structure and placement of the engine-passenger module forward for improved weight distribution, the Airflow was a major engineering development. These concepts rapidly became the accepted norm for almost all front engine-rear drive cars in the entire industry. Unfortunately for Chrysler, the Airflow styling concept was not widely accepted. Airflow aficionados claim that it was a splendid car to drive.

1934 Citroen 7A (France) (photo no. 48)
6 cylinder 80 cu in. (1303 cc) 32 hp engine (7A)
6 cylinder 175 cu in. (2866 cc) 76 hp engine (15CV)

Perhaps one of the most successful engineering concepts in automotive history, this design quickly became France's best selling car. Known as the Traction Avant, after its superb front wheel drive, all engine accessories, radiator, drive train and front wheels were assembled as a unit and plugged into the frame. A more intelligent approach is difficult to imagine. To create the car, Andre Citroen engaged Andre Lefebvre, an aeronautical engineer, as project chief of a distinguished staff of engineers and designers including Maurice Sainturat for the engine design and Flaminio Bertoni as stylist. This Citroen masterpiece set the pattern for front wheel drive automobiles. Years later, the Japanese auto firms of Nissan, Mitsubishi Motor Corporation and Nippon Seiko KK, acquired licensing rights to use the Citroen front wheel drive patents. The engine module was not the only thing which lent merit to the car. It was the first mass produced European car using all-steel *monocoque* (one shell) body construction, wherein the body shared the burden of absorbing shock and twisting with the frame (unitized body and frame). The lowering of the body and frame, made possible by the absence of a propeller (drive) shaft to the rear wheels, made it possible for Bertoni to reduce the frontal area by 20% compared to previous Citroen designs, achieving a significant reduction of aerodynamic drag. Torsion bar suspension on all four wheels, hydraulic brakes and the low center of gravity gave it excellent handling characteristics.

In one of the larger production runs in automotive history, 760,000 were built from 1934 to the last Traction Avant off the production lines in 1957.

1936 Cord 810 (United States) (photo no. 49)
V-8 cylinder Lycoming 289 cu in. (4735 cc) engine.

The Cord Corporation was aware of the shortcomings of its earlier L-29. Not just its unfortunate 1929 timing influenced its demise. One major flaw was the arrangement for getting power to the front wheels. The L-29 configuration placed the transmission, attached to the front of the engine, aft of the front axle centerline, pushing these components and the passenger module too far back. Also, the suspension design resulted in excessive weight of the parts which respond to road contours (unsprung weight), an undesirable condition. The Cord group commissioned their designer, Gordon Buehrig, and their engineers to develop an entirely new vehicle, free of past limitations and thoroughly contemporary in engineering and design.

The Cord 810 was a beautiful automobile, albeit a maverick. Though the engine-passenger module was placed forward, thanks to a transmission gear box ahead of the front wheel centerline, its design, while charming, was not in vogue with the times. The car's distinctive hood (earning it the sobriquet "coffin"), prominent fenders and overall body style departed from look of the mid-thirties. By the end of 1937, both the Cord 810 and its successor, the 812 of 1937 went the way of the L-29, with about 3,000 810's and 812's produced.

A semblance of the Cord survived for a while as the dies which formed the body parts were sold to Graham and to Hupmobile. Both companies attempted to fuse the distinct Cord lines with their own ideas of styling, without notable success.

As related in this book, the Cord 810 and 812 transmissions were resurrected as a temporary salvation for the Tucker '48. Ironically, the first pilot run of Cord 810s in 1936 had been rushed to completion and shown without transmissions in place, these rather vital parts not yet completed.

1936 Dubonnet Dolphin (France) (photo no. 50)

V-8 cylinder Matford 225 cu in. (3695 cc) 72 hp engine.

Andre Dubonnet, a scion of the French family famous for its wines, was an automobile enthusiast with a passion for inventing advanced automotive ideas. In the early 30s, he showed a Dubonnet sedan at the Paris salon. The body sat on a 129 in. (328 cm) chassis powered by a six cylinder Hispano-Suiza engine with a radically new suspension system. A complicated mechanism of hydraulic cylinders and pistons dampened road shock and rebound on each independent wheel. General Motors acquired the patent rights for this concept and, in much altered form, introduced it on the front wheels of 1934 Chevrolets.

Dubonnet continued his quest for new ideas and in 1934 developed the Dubonnet Dolphin, styled by Hibbard of the United States in partnership with Fernandez of France. This car stands alone in its era. It used the idea of a car with its engine aft of the passengers, but sensibly placed the weight of the engine and its transmission forward of the rear axle centerline. Immediately aft of the transmission, the differential was fixed, delivering power to the rear wheels through swinging axles on universal joints. The engine was a French Matford V-8 (the French version of the Ford V-8), the transmission an electromagnetic Cotal four-speed. Dubonnet included his novel suspension ideas in the independent springing. His aim was the ideal European vehicle, small, lithe, well-balanced and well suited to the rough and curving roads of the countryside.

At the Montlhery race track in France, the Dolphin's 72 hp Matford engine was pitted against a standard Ford V-8 of 80 hp. The Dolphin's best lap was 108 mph (173 kph); the Ford's, 82 mph (132 kph). Dubonnet brought his dream car to the United States in the spring of 1936 and demonstrated it to Henry Ford. It was rejected. The Dolphin next went to the General Motors test track where it no doubt raised eyebrows, but garnered no contracts. In spite of its performance and numerous features, only one was ever built. The car faded into obscurity, but its ideas did not. The Dubonnet was a classic mid-engine car; the Pontiac Fiero of today is only one of a number of mid-engined makes on the road fifty years later. However, as William J. Lewis of the Society of Automotive Historians states, "The front opening door (for the driver and the front seat passenger) was a strike against the otherwise technically successful streamliner."

1936 Lincoln Zephyr (United States) (photo no. 51)
V-12 cylinder 268 cu in. (4392 cc) 110 hp engine.

Appropriate to its name, the flowing lines of the Zephyr were a refreshing breeze in comparison to many cars of its time. Designed by John Tjaarda and Howard Bonbright of Briggs Manufacturing and E.T. Gregorie of the Ford Motor Company, the Zephyr was based on the Tjaarda 1933 rear engine streamliner (shown to the public in 1934). Technical problems and a consumer poll are said to be the reasons for changing engine location to the front. It is difficult to believe that the Ford Motor Company engineers were unable to surmount design problems inherent in rear engine placement (a technique already perfected elsewhere). Perhaps the poll, indicating consumer preference for front engine location, swayed management (which shows what happens when consumers are asked to choose something without being given necessary technical knowledge). The Zephyr was a success, but the inadequate torque and maintenance problems of the new engine designed to propel it tarnished its reputation.

1938 Cadillac Sixty Special (United States) (photo no. 52)
V-8 cylinder 348 cu in. (5702 cc) engine.

This significant departure from heavy, rounded window structure, thick pillars and ungraceful contours was designed by William L. Mitchell in 1937 when he was twenty-five years old (Mitchell went on to succeed Harley J. Earl as vice president in charge of styling at General Motors in 1958).

A distinct lack of running boards, gracefully blended lines, sharp pillar cross-sections of reduced widths, small radii of window trim, and increased vision provided by a windshield slanted 39 degrees which went almost to the top of the body, distinguished this relatively low-slung beauty. The Cadillac Sixty Special was only 65 inches (156 cm) high; quite radical for American cars of that time.

Cadillac heralded it as being unusual, even to its being the fastest of all eight cylinder automobiles. Ads proclaimed it "The best performing American stock car." However, even Cadillac did not appreciate the impact it would have. Its slender roof supports and lean lines launched the hard top look that was eventually emulated almost universally by the auto industry. Most post WWII cars produced by Detroit were not as advanced in style as the 1938 Sixty Special.

An overriding concern of the designers behind these automobiles was the quest for change, for improvement, for a different and better car. Some demonstrated a styling thrust into the future to simplify lines and fill in revealing gaps. Others sought a lower profile and better traction by virtue of front wheel drive. A number strove for better aerodynamic shape to better cleave the air and improve fuel economy. Many presented shapes for purely styling considerations, perhaps to stimulate buyer response.

A few, like Tucker, attempted to accomplish engineering advances, superior layout and a more aesthetic form, all in one grand magnum opus. There is little doubt that some of these designs established meaningful advances, while the flaws which doomed a few to extinction were obvious.

Not to be overlooked were the excursions into pure automotive artistry for its own sake, probably best exemplified by British and Italian efforts. A classic example is the 1938 Jaguar SS 100 (photo no. 54). This exciting creation made no pretense of being a representative wave of the future. Its huge headlights, radiator set well back within imposing wheels and exposure to the elements for the driver presented a picture of motoring delight. The SS 100 is one of the best examples of a sheer sport automobile contrived in the period before World War II.

Up to the beginning of World War II, in 1939, the roll call of new ideas in the automobile industry was impressive. The engineering, configuration and styling ideas of Jaray, Ledwinka, Porsche, Citroen, Wright, Tjaarda, Gregorie, Dietrich, Mitchell and many others firmly set in place a beacon with which to light the future. To an industry which had become a major component in world economics, they gave the basic elements for the exciting developments of the years ahead. In a little over a half century of their efforts, the automobile has evolved from a carriage with the horse replaced by a chugging, primitive, internal combustion engine, to a glistening form in which driver and passengers can ride swiftly in comfort.

APPENDIX II

PROFILES

J. Gordon Lippincott: "There is no such thing as 'retirement.' I simply stopped working for money and work for fun. I have a great terrace overlooking the sea in the Florida Keys, but rarely have time to sit down and soak up the sun!"

Tucker P. Madawick: "I found the ultimate life in Florida with important things like sailing the Gulf with my wife, Patricia, and collecting classic cars. Still dabbling in the exciting world of design."

Budd Steinhilber: After seventeen years as a partner of Read Viemeister's in Ohio, Budd went to San Francisco and formed his own design firm, first with Gene Tepper and then with Barry Deutsch. He retired in early 1988, and now lives in Hawaii where he maintains an individual design practice.

Alex Tremulis: Went to American Motors after the Tucker demise to apply for a job. The chief engineer there said, "You are the eighth designer who claims to have designed the Tucker." When the engineer looked at Alex's portfolio, he laughed and apologized. "The other seven," he said, "were bums."

Read Viemeister: Maintains his design consultancy in Yellow Springs, Ohio, in friendly competition with his son, Tucker, a designer in New York.

Hal Bergstrom: The lone non-survivor of the Lippincott Tucker crew, is no longer around to tell us of his part of the story. He was a fine gentleman, a consummate designer and memorable companion.

ABOUT THE AUTHOR

Philip S. Egan is an industrial designer, artist and writer. He was one of the designers of the Tucker '48, first as a member of the New York based Lippincott and Margulies design team, and later as assistant to the chief stylist of the Tucker Corporation, Alex S. Tremulis.

After the Tucker venture concluded, Egan went on to design numerous automotive, aircraft and astronomical products. He was the designer of the Sears/Replac Debonnaire fiberglass sports car. For Sears and Cushman Motors, Egan designed several motor scooters, and light trailers for Sears and Dunbar Kapple. He has designed consumer items ranging from tractors and sewing machines to furniture and toys. His work has also included designing residential and medical laboratory interiors, and devices for the hearing and speaking handicapped. Philip Egan is a member of the Industrial Designers Society of America.

Design and Destiny is his fourth book. His previous books, *Space for Everyone*, on space travel and astronomy written soon after the Sputnik and Mercury flights; *Where in the World*, a broad view of geography; and *Know Your Rivers of the World*, were published by Rand McNally & Company.

In addition to writing, Egan is a design consultant in northern California, where he lives with his wife, Virginia. A current project is a limited edition series of watercolors of notable automobiles and aircraft.

The Tucker Automobile Club of America was founded in 1972 and is still active today with over 500 loyal Tucker admirers.

Those readers wishing more information about the Tucker Automobile Club of America can write to Richard E. Jones, 315 Arora Blvd., Orange Park, Florida 32073.

ON THE MARK Publications would like to acknowledge those people without whose help this book could not have been published.
Peter Arnold
Arnold M. Cowan
Wayne and Polly Hendriks
Henry D. Ives
Christine Lowentrout
Lee Vibber

INDEX